T0134081

Physician Integration & Alignment

IPA, PHO, ACOs, and Beyond

Physician Integration & Alignment

IPA, PHO, ACOs, and Beyond

Maria K. Todd

CRC Press is an imprint of the
Taylor & Francis Group, an **informa** business
A PRODUCTIVITY PRESS BOOK

CRC Press
Taylor & Francis Group
6000 Broken Sound Parkway NW, Suite 300
Boca Raton, FL 33487-2742

Printed in the United States of America on acid-free paper
Version Date: 2012928

International Standard Book Number: 978-1-4398-1308-9 (Hardback)

Library of Congress Cataloging-in-Publication Data

Todd, Maria K.
Physician integration & alignment : IPA, PHO, ACOS and beyond / Maria K. Todd.
p. ; cm.
Physician integration and alignment
Includes bibliographical references and index.
ISBN 978-1-4398-1308-9 (hardcover : alk. paper)
I. Title. II. Title: Physician integration and alignment.
[DNLM: 1. Delivery of Health Care, Integrated--organization & administration--United States. 2. Accountable Care Organizations--organization & administration--United States. 3. Financial Management--organization & administration--United States. 4. Hospital-Physician Joint Ventures--organization & administration--United States. 5. Independent Practice Associations--organization & administration--United States. W 84 AA1]

362.1068--dc23 2012021346

Visit the Taylor & Francis Web site at
http://www.taylorandfrancis.com

and the CRC Press Web site at
http://www.crcpress.com

Contents

SECTION IV Appendices

Acknowledgments

As a consultant with more than thirty years of experience in healthcare, and the unique perspectives coming from the clinical, health plan, hospital administration, group practice management, and health law paralegal experience, I see things from a different perspective than my colleagues. This is a good thing—and at the same time a challenge. The good thing is that I am able to see the "big picture." The challenge is that my nature is to be a critical thinker, and to not be shy when someone asks me what I think. One cannot be an author, a speaker, or a consultant without having the courage of one's convictions and confidence. To the casual observer, I am often viewed as, well, shall we say, "difficult." Those who have made the investment of their time, skill, and patience to get to know me, to challenge me, to exchange views and insights with me, and come away knowing that I am a skilled professional with high standards—those are the ones I wish to acknowledge.

I would like to thank all those people who have helped me in the development of this book. I found it inspiring that friends, past clients, and strangers have been willing to give up their valuable time to review and comment on the early drafts of the book. Some of you reined me in, others helped me develop concepts more fully, and still others just simply provided a different perspective than the one I held, which I ultimately used to dig deeper and share more in the chapters.

What I have learned over the years from friends, colleagues, and clients is that there are myriad choices available in the development of provider organizations. It is similar to building houses. Most homes have bedrooms, bathrooms, a kitchen, some space to receive guests, and a place to relax. Houses can range from a rustic, simple cabin in the woods, to an elaborate Victorian mansion, to a functional Craftsman bungalow, and everything in between. Yet, for the most part, we still call it our "house." When someone invites you over for a visit, they don't say, "Come on over to my Victorian mansion and we'll roast a hog in the backyard." They don't say, "Come over to my mid-century, modern split-level ranch and we'll make fondue." And, they don't invite you over to their rustic cabin in the woods for a Vegas-style buffet. What they say is "come on over to my house." So you put the address into your trusted GPS device, navigate your way to the address, and when you get there, what happens? You get the first glimpse of what they call "their house." Most likely, it is not a carbon copy of anyone else's house that you have ever seen, unless the builder used an architectural plan kit, or based it on a house they built for someone else because they knew how to build only that one design.

Building IPAs, PHOs, MSOs, and ACOs is much the same. I have to thank my colleagues for sharing with me their "house design project" notebooks along the way, and my clients for providing me with a wide range of design options that they wanted for their "houses." As I read the acknowledgments of my first work on this subject—*IPA, PHO, MSO Development Strategies: Building Successful Provider Alliances*—without reiterating all the acknowledgments included there, I wish to thank them all again, as without that first work, this revision would have been much more difficult to produce.

To Maria Scouros, MD; Pat Chinn, MD; Calvin Wong, MD; Lockwood Young, MD; Herb Chinn, MD; Tim Owen, MD; Dick De Journett, MD; Gene Nobles, MD; the late Claude Ledes, MD; Bellur Ramanath, MD; Steve Combs, OD; Marty Sikorski, OD; Dan Shea, PsyD; David Goldstein, MD; Mark Rhine, MD; Howard Corren, MD; Richard Patt, MD; Bruce Hayward, DO; Leslie Corcoran, MD; Bill Solomon, MD; Gerry Kirshenhaum, MD; Bob Levy, MD; Dr. Jim McKeown; Steve Arter; Albert Hollaway; Barry Drexler, LMT; Seneca Pelton, LMT; and Ellen Stewart, Gregory Piche, David Samuels, and others: I have learned much from you, and I bring with me the lessons you have taught me as I teach others.

The team behind the writing of this book deserves much of the praise from me. First, the folks at CRC–Productivity Press, namely, Kristine Rynne-Mednansky, one of my closest professional associates and senior editor. She has had the patience to let me work around life's little bumps in the road. She has moved due dates, and acted as a sounding board when I decided not to place too much confidence in or emphasis on ACOs. Also, my fellow healthcare futurist and esteemed medical economist, Jeff Bauer, PhD, who contributed the new title to the book when I became uncomfortable with the working title that Kristine and I originally contemplated, given where the integration in healthcare has gone, and concurred with me when I expressed my reservations about ACOs and wanted to lessen its importance in a book I felt would live longer than many yet-to-be-declared ill-fated physician alignment and integration projects.

What is different in this book from the original work of 1996 is the acknowledgment and thanks to my connections on LinkedIn, Twitter, Quora, and followers of my blogs, both those well known to me and those who share groups with me and who have generously given of their time to act as a sounding board and share insights on best practices, lessons learned, and different perspectives.

Finally, to my husband Alan, my friend and partner of over twelve years: I am done … with this book. (Look out Kristine, he might throw something at both of us!) Through five books in three years, you have been incredibly patient and understanding. You have taken over many of the household tasks, spent many evenings alone as I traveled on business, and, most of the time, held your tongue regarding my long hours and busy travel schedule.

About the Author

Maria K. Todd has been involved in the healthcare industry for most of her working life. She has maintained an independent consulting practice since 1986 and has offered guidance to thousands of clients worldwide in the domains of managed care, healthcare revenue cycle, hospital and medical group administration, physician employment contracting, organizational development, medical tourism, healthcare benefits management, and value-based purchasing.

With her multifocal background and education as a healthcare business administrator, health law paralegal, surgical nurse, HMO provider relations coordinator and certified mediator, and a licensed insurance producer, she has expertise in niche areas such as full-risk capitation, managed-care contracting, and negotiating on behalf of payers, providers, and employers. She brings a wealth of specialized knowledge to the development, implementation, and operation of IPAs, PHOs, MSOs, and other integrated health delivery systems. She draws upon this extensive experience to bring value to physician groups, boards of directors, and ministries of health in more than thirty countries as a teacher, author, speaker, and consultant.

In 2009, she filed for and, after four denials and appeals, successfully prevailed in registering a trademark for a new term of art in the industry for the Globally Integrated Health Delivery System®. The U.S. Patent and Trademark Office accepted her application in August 2010, granting trademark registration to define the term to describe a game-changing organizational structure and function that converges integrated health delivery, key principles of managed care, patient access, care continuity, electronic health information technology for global electronic exchange of health information to improve quality of healthcare, and health travel/medical tourism. The model is now in operation as Mercury Healthcare International, and is positioned to accommodate the healthcare needs of a flatter, more mobile, global society.

Todd's previous consulting projects have resulted in the launch and implementation of more than 150 successful IPAs, PHOs, and MSOs in medical, surgical, behavioral health, complementary and alternative medicine, and ancillary service providers. She has provided expert testimony and supported forensic economists in litigation on failed IPAs, PHOs, and MSOs projects; provided support to private equity investors and market analysts; collaborated with other business consultants and attorneys to help develop private placement memoranda for IPAs, PHOs, and MSOs; provided expert testimony for antitrust litigation brought by "locked-out" IPA and PHO providers; and mediated disputes between payers and provider organizations, and between hospitals and physicians developing PHOs.

A frequent speaker in the United States and abroad, and a former member of the McGraw-Hill Healthcare Education Group's seminar leaders, Todd speaks before numerous state, national, and local organizations; government agencies; hospitals; and provider groups. She is available for in-house training sessions on a variety of topics related to managed healthcare, capitation, provider and network contracting, and integrated delivery system development and management. You may reach her at her office in Denver, Colorado, at (800) 727-4160, or via email at maria@mariatodd.com.

Introduction

Early in my education, I had an inspiring professor who introduced me to the principles of organizational development (OD). He lit a fire in me that never died. I reflect on what influence that course had on how I lead, consult, and manage others. At the time I took that course, I was already a beginning consultant. I had been consulting with physicians for about three years (1980s).

While most of my classmates had not yet entered the business world, I was able to read the textbook, apply it to what I was doing day in and day out, and bring reality and recent examples to class discussions and exercises as we progressed through the book, *An Experiential Approach to Organizational Development.** What I loved most about the book and the course was that it differed from most OD texts in providing us with a strong conceptual framework, descriptions of the then-current, state-of-the-art approaches, techniques, and methodologies for implementing OD programs; and used learn-by-doing behavioral skill simulations for each major stage of an OD program.

All of my integrated health delivery system projects came after taking this course and putting what I learned into practice. I followed with additional study and practice through the American Society of Training and Development (ASTD), which further influenced my perspectives and my teaching and presentation styles … and still does to this day.

Too many IPA, PHO, MSO, ACO, and integrated health delivery system projects are undertaken without adequate capitalization, infrastructure, or the leadership skills necessary to guide the stakeholders through successful completion of the first year of operation, contracting, USP development, process improvement, outcomes measurement, and continuous quality improvement. Hopefully this book will shorten the learning curve, save some money on consulting education, save time, and enable the reader to appreciate what goes into the development of a successful provider alliance.

As you read through this book, you will read what may appear to be sarcasm. In some cases, you will see clearly that I am being sarcastic or facetious about something. In other cases, what sounds like sarcasm is actually pure, unconcealed passion, frustration, and incredulity at what I see around me in ill-fated physician alignment and integration projects that are mismanaged, lack a unique selling point, offer no value to the marketplace, solve no problems, do not align stakeholders or answer the WIIFM (What's In It For Me) question, or were the product of the pressure to build one because everyone else is doing it.

I have broken the book into four sections, starting with an introduction of the key terms (Section I, "Introduction to Provider Organizations"), namely, alignment and integration, and then present short chapters to introduce each of the four acronyms.

* Harvey, Donald F., and Brown, Donald R., *An Experiential Approach to Organization Development, 3rd edition*, Upper Saddle River, NJ: Prentice Hall, 1988.

The way I choose to present my point of view is to develop an analogy between the development of these healthcare provider organizations and building a custom house. Sigmund Freud said, "Analogies prove nothing, that is true. But they can make one feel more at home." My writing style, for those of you who have never read any of my previous books, is very casual, relaxed, and in the tone of a fireside chat over a glass of wine or a cup of cappuccino. It works for me; and I truly hope it works for you.

In Chapter 1, "The Goals and Objectives of Physician Alignment and Integration: Form Follows Function," I start with the analogy of building a custom home and share a personal example of how my husband and I differ in taste, goals, and personal style, yet choose to be together, much the same as a group of physicians forming a provider organization of some sort. First, they need to determine what they need in order to have alignment with their personal and professional goals and objects, work habits, style, preference, and cultural values. Then they need to find others who they are able to align with, and build the organization so that the integration functionality aligns with them as stakeholders and the corporate form of the organization. The chapter is a quick read, but it sets the tone for what lies ahead.

Chapters 2 through 5 identify, at a 30,000-foot level, the IPA, PHO, MSO, and ACO. I hold the opinion that after building more than 150 of them it really is not about what you call it, and if you have seen one of them, you have seen all of them. As analogies are reasoning or explaining from parallel cases, it is easy to understand why I chose the analogy of the custom home project.

Section II, "Integrated Health Delivery System Development" (Chapters 6 through 14), is where I believe the primary value of the book is stored. These nine chapters explain processes, offer checklists, lessons learned, and templates and models that will save you both time and expense. This is a gift that will pay for the cost of the book and the time invested to read it many times over.

My philosophy is "see one, do one, teach many." On my death, when they scatter my ashes, I hope my loved ones say those words as the ashes make their way to settle wherever they will.

Most consultants argue that I give too much away in these chapters. I hold that there are plenty of consulting projects to go around and that if the client is better prepared, the project has a greater chance of success.

In Chapter 6, "Corporate Form: Myriad Choices," I provide checklists and guidance for your initial mission as a committee, an agenda, and an introduction to what it takes to produce steerage of these organizations. The role does not really change, regardless of whether or not it will evolve into an IPA, PHO, MSO, or ACO. The tasks are essentially the same.

In Chapter 7, "The Steering Committee Gets Busy: Step-by-Step Instructions for What to Do and How to Do It," I have done my best to describe what others have done and the elements that you must consider together with competent legal and accounting personnel, of which I am neither.

In Chapter 8, "Guidance for the Utilization Management and Quality Improvement Steering Committees," I present the basic responsibilities for these two committees. The tasks undertaken and oversight in these areas will be vital to the success of creating value for both consumers and stakeholders.

In Chapter 9, "Network Financial Management: The Intersection of Finance, Utilization Management and Capitated Risk Management," I attempt to bring the "suits" (the administrators and financial folks) and the "lab coats" (the clinicians) to the table to play nice. If the three committees cannot lead the organization through the quagmire of aligning reimbursement with clinical delivery, it is over.

In Chapter 10, "Provider Organization Credentialing and Privileging," I attempt to explain the vital processes of quality "assurance" or "reassurance," which carries with it both ostensible and vicarious professional liabilities. Most physicians actually have little appreciation for what goes into medical staff services, credentialing processes, and privileging decisions because it often happens behind the scenes at the hospital or the managed care plan.

In Chapter 11, "The Credentialing Committee's Assignment: What to Do and How to Do It," I offer a high-level summary of what must be done by this committee and those under their oversight. Chances are, the physicians will not personally be performing the processes involved in primary source verification; but without an awareness and appreciation for the detail and labor involved, and the complexity of process, they risk underestimating the costs of the process and the need for expertise that comes from qualified consultants and detail-oriented employees who will carry out the labor associated with the tasks.

In Chapter 12, "Antitrust Compliance Task Force: Understanding Antitrust Concerns for Provider Networks," I essentially reiterate the guidance from the U.S. Department of Justice (DOJ) and the Federal Trade Commission (FTC). In the years since the guidelines were first published, I have witnessed many organizations start out legal and end up prosecuted. (None were my clients, however! Whew!) I have agonized as colleagues who were their consultants were also involved in the legal actions because they gave bad advice, sought to tread into gray areas, and perhaps, out of ignorance, recklessly stepped beyond their scope. The end results can be devastating. The need for competent legal counsel cannot be overstated.

In Chapter 13, "Business Plan Development," I offer an outline that should get you started but you will need the right financial and legal counsel to bring the project to a point where it can serve as a proper road map for organization development and communication of the vision to stakeholders.

In Chapter 14, "Guidance for the IT Committee," I figure that you all get that you must have some sort of electronic health record that offers security, medical records repository, and meets HIPAA (Health Insurance Portability and Accountability Act) guidelines. I also assume you get that you will need a website, and some reports, so I did not belabor those perfunctory concerns. Instead, I take the chapter in the direction of business intelligence, and offer an analogy to the significance of healthcare information technology and systems as a parallel to the "Circle of Willis" in functionality and sustainability.

Section III, "Business Development: Contracting and Marketing," includes several chapters to take readers beyond the start-up and creation of the provider organization and into the functional areas of business development.

In his best-selling book *The Seven Habits of Highly Effective People*, author Stephen R. Covey says, "begin with the end in mind."

In Chapter 15, "Contracting with Payer Organizations," I take readers beyond typical HMO and PPO arrangements and into other likely arrangements for the road ahead, including anticipated changes in reform and beyond into value-based purchasing, direct contracting with self-funded employers and labor unions, and purchasing coalitions.

I continue with Chapter 16, "Contracting for Capitation and Bundled Service Arrangements," where I address some of the technical aspects of contracting for reimbursement other than fee-for-service methods. Here again, there is so much that could be addressed in this chapter, but I choose to keep it high level, and suggest that if further skills-building is needed, that you take one of my classes or obtain a copy of my *Managed Care Contracting Handbook, 2nd edition,* which develops this particular area much more deeply.

In Chapter 17, "Understanding Capitation Performance Guarantees," I present some of the lessons learned in the 1990s when many organizations went bankrupt because of poorly analyzed contracts and a lack of appreciation for the potential risk of IPA, PHO, or MSO capitation.

In Chapter 18, I continue the topic of capitation and bundled payments with "Considerations for Reinsurance Purchases for the Integrated Health Delivery System." For success with capitation and financial risk assumption, you will need competent counsel from underwriters, brokers, and actuaries. You will also need a budget as the policy for reinsurance will be expensive, but compared to the financial protections it offers, the trade-off is small. The broker and the underwriter and the actuary will work together to offer competent counsel to attempt to estimate the risk within a reasonable margin of error so that you do not purchase inadequate coverage, and do not pay too much for coverage you may not need.

In Chapter 19, "Opportunities in Delegated Utilization Management and Claims Management for the MSO," I offer some insight into ways that provider organizations can earn revenues beyond treating patients. But I also attempt to highlight some of the complexities involved so that the reader gains an appreciation for the technical aspects of these often underappreciated and misunderstood processes and workflows that happen behind the scenes at the third-party administrator or health plan headquarters—or offshore in business process organizations (BPOs) in less-expensive labor markets.

In Chapter 20, "Beyond Traditional HMO and PPO Contracts: Direct Contracting with Employer-Sponsored Health Benefit ERISA Plans," I wrap up with the grand finale for what I believe is the next phase, beyond government-driven healthcare reimbursement reform and insurance reform to market-driven reform and value-based purchasing. Early adopters have already been engaged in this for a solid twenty years. My first projects, still successful, were in 1993 and 1995.

In the final section of the book, Section IV, I include three appendices. In Appendix A, the "Volunteer Committee Survey Form" template will get you started in rallying the troops and setting the culture of the organization. None of the groups I have ever worked with has successfully launched by allowing some members to toil on organizational development, while others simply write a check for their share. Engagement is key to alignment and integration.

Appendix B is a glimpse of what is needed in documents if you elect to form a Limited Liability Corporation (LLC). The "Sample LLC Document Set" is not a template for filing your papers. Instead, the objective is to guide your thoughts into the co-authorship of these required documents together with your attorney. Again, it is like reviewing the final plans for the house before they are submitted to the city and the county for permits.

The final appendix, Appendix C, "How to Hire the Right Consultants," will hopefully save you some time, frustration, and aggravation and will lead you to being a better consumer of consulting services, which will hopefully lead to adequate, appropriate, and competent counsel to meet your organizational goals and objectives.

The guidance that I have attempted to convey in this book is used to underpin these four types of organizations, from concept to business sustainability and success. They are not offered as sacrosanct principles that must be employed in all situations; rather, the aim of the book is to offer an outline framework within which each group of physicians or ancillary or allied health professionals and their chosen partner facilities can understand the context of a situation and then develop a solution that is appropriate.

I hope you accept my ideas in the spirit with which they are offered. Although my intention is always to offer ways and means, not "must" and "should," this is difficult when certain ideals and beliefs are felt with passion. While parts of the book may appear directive in nature, if you ever feel that the style or content is offering the only way to do something, feel free to consign it to the recycle pile.

Section I

Introduction to Provider Organizations

1 The Goals and Objectives of Physician Alignment and Integration

Form Follows Function

Similar to building a custom home, there are design considerations that work for some inhabitants and not for others. Construction costs vary due to dozens of factors in a custom home design. The exterior materials, the number of floors, the site, the interior finishes, the geographical area of the country, the complexity of the design, and the structural and mechanical systems used are only a few of the factors that influence the cost of constructing a custom home.

A structural engineer is required for homes in most parts of the country. An interior designer may, or may not, be required to finish one's dream home. However, most people find the services of an interior designer to be invaluable during the design process. For larger luxury custom homes, the need for careful integration of everything from fabrics to floors usually requires the services of an experienced and sensitive interior designer. Landscape architects, lighting designers, mechanical engineers, surveyors, soils engineers, pool consultants, audio/visual consultants, and kitchen designers are some of the other consultants often brought onto the team on a luxury custom home. Your architect acts as the team leader, orchestrating the consultants and the builder to achieve the maximum in creativity and enthusiasm from everyone involved in your home design.

Now let's move that analogy over to building a "custom home" for physicians—one in which they will feel comfortable, relax, grow their business, thrive, enjoy their "neighbors" (professional colleagues), and rely on the local healthcare facilities in times of medical necessity.

What aligns the homeowner and the architect? I believe it is validation and keen listening skills and communication going two ways. I want the things I want. There is no right or wrong: it is my house, I want the laundry on the same floor as the master bedroom, I want a dual fuel oven, I want open spaces and little nooks, places to sit and watch the snow, view the mountains, and read a book.

I want sustainable materials, from local sources whenever possible. I want energy-efficient lighting and natural light sources. We both enjoy electronics and technology, but I want electronics to seamlessly fit in a room without wires—trip hazards for me, injury hazards for cats that like to chew, and falling hazards for my bad knees.

I want efficiency, in that when I bring home the groceries, I do not want to have to lug them very far to put them away. I want space in my kitchen so we can entertain while we cook; I want room so that both my husband and I can work in the kitchen on different dishes without getting in each other's way.

He wants places for a large collection of books, all the cooking and food preparation gadgets, and high-quality cookware, knives, and tools. He wants room for the Weber® gas grill, the Weber Smokey Joe®, the big charcoal kettle grill, and his Big Green Egg®.

We both want walls where he can display his extensive collection of platinum prints, orotones, and photogravures by Edward S. Curtis without subjecting them to changes in temperature and moisture on exterior walls, but also walls without damaging UV and high-altitude sunlight. He wants rooms large enough to use our original arts-and-crafts rocker, Morris chairs, and other furniture from Stickley, Roycroft, and other craftsmen.

We also love pottery from Russell Wright and mid-century modern era design, and we have pieces by Florence Knoll, Marcel Breuer, and others that we want to integrate into the house because we use them—we don't just have them for show.

I like to entertain; I have Lenox® china and all the completer pieces that need to be stored when not in use. I have a complete set of Reidl glasses for every kind of wine and whiskey and cordial. I do not have Bronco's® commemorative glasses, but I do have depression glass handed down from Alan's family, pieces I have collected throughout the years that I like to use as serving pieces. I like to entertain at the spur of the moment, so they need to be stored, yet accessible when I need them, without traipsing off to the basement or out to the garage to get them.

Alan likes red wines; I get migraines. I like white wines, which he argues haven't the body and robustness he enjoys. I argue that I hate pain and no amount of tannin is worth the threat of a "master blaster." I like single malt scotch; he likes tequilas and is known for his superb margaritas. Mine takes a simple neat glass, a cube, a splash. His takes a blender, sticky juices, limes, rind, peelers, knives, odd-shaped glasses that take up space and makes a real mess in the kitchen.

We share our home with two cats, Jack, age 4, a social, Labrador-like natured classic red tiger tabby, and Grace, a seven-year-old, street-savvy Siamese that talks a lot and likes *mu, gai,* and *kung* (those are Thai words for pork, chicken, and shrimp). Grace also drinks coffee and eats Mexican food, lasagna, and greens from my window garden. Jack likes to watch water flow and to curl up on our laps or desks, or counters—wherever we are. With as much time as he has spent on my desk listening to me mutter while I review managed care agreements, he probably has jaded impressions of many HMOs (health maintenance organizations) and PPOs (preferred provider organizations) at this point that he will never reveal. With as many presentations as he has "helped" me develop, I am betting he is also a master at PowerPoint. Grace, in comparison, likes nooks, crannies, privacy, and quiet. The only thing she likes to manipulate … err … manage, is Alan. They will hopefully be with us for years, so we need a design plan that works for them also.

I get cold easily. So when we entertain outside, I want one of those gas-fueled "heater thingies" that looks like a big stainless steel lamp, and a wood-burning chiminea that crackles and smells of wood smoke. I have fibromyalgia, so I want/need a hot tub or spa … with lots of jets!

We both love tending a spring and summer garden and grow heirloom tomatoes, sweet banana peppers, chili peppers, squash, and lots of fresh herbs. With my knees, we need raised beds. That means there must be sunlight in strategic places at strategic hours.

Over the years, we have made do with home designs that were built for other people—tract homes, mountain homes, a craftsman bungalow built in 1928, and apartment homes. We tried to make our lives fit into homes that were designed as part of a Las Vegas covenant-controlled community … like "Stepford," a sprawling mountain custom home built in the 1960s on the face of Pike's Peak, accessed by a steep dirt path, and a 1928 Craftsman bungalow near downtown Denver in an area that is experiencing gentrification—but none have really been "aligned" with our needs and preferences. As a result, it has been difficult to "integrate" our lifestyle and design style into something that flows, that works for us and with us. Instead, while we love and respect one another, we have had frustration, tension, and aggravation, although not necessarily with one another, but instead, with the feeling that we are attempting to make our square-peg lives fit into someone else's round holes.

We have come to a decision that we need a custom-designed home that "fits," and we now are faced with two choices: redesign and remodel where we live, or start fresh with a custom design. In either case, we will need the architect, engineers, designers, a budget, plans, a project timeline, permits to satisfy regulatory requirements, accounting of costs for assistance, materials and labor, and cash to adequately see completion of the project, whether it comes from cash on hand or borrowed from a lender. If from a lender, there must be a repayment schedule, with a return on the investment the lender has made, a security agreement, and so forth.

The point is this: no two houses are going to be alike, even though we might call both of them houses. The alignment of the inhabitants or, in the case of a provider organization, is paramount as a step toward the ultimate goal of integration. The job of the architect and the steering committee is to achieve effective integration, which goes beyond alignment, and is achieved when the individual components of the house operate as a fully interconnected unit.

The integration between the homeowners and their "stuff" is one kind of integration. The integration of the homeowners and their neighborhood elements—for example, the local healthcare facility—is another type of integration. The words may be the same, but the context is completely different.

In a healthcare context, physician alignment refers to the linking of an integrated health delivery system's organizational goals with the physicians' personal goals. To do this requires a common understanding of the purposes and goals of the organization, and consistency between every objective and plan right down to effectively addressing the most challenging obstacle for each physician's ultimate engagement: answering "WIIFM*?" to align with their business, professional, investment, and personal goals and objectives.

Corporations, hospitals, and physicians and allied health professionals often find it impossible to bridge the gap between themselves because of differences in objectives, culture, and incentives, and a mutual ignorance for the other group's body

* WIIFM: What's In It For Me?

of knowledge and how they process and perceive the world around them. This rift generally results in the expensive business development of an Independent Practice Association (IPA), Physician Hospital Organization (PHO), Management Services Organization (MSO), Accountable Care Organization (ACO), Clinic Without Walls, or other legally, economically, or clinically integrated health delivery system that does not provide adequate return on investment to survive.

It is not unusual for hospitals, corporate administrators, physicians, and allied health professionals (stakeholders) within an organization to experience conflict and in-fighting as lack of mutual understanding and the failure to produce desired results leads to blaming and mistrust. The search for alignment often includes efforts to establish trust between these two groups and a mechanism for consensus decision making. Often, it is because people just "expect" others to be professional and maintain a certain integrity, and they fail to plan for these challenges in the initial strategic development of the organization and deploy processes to account for them and support the stakeholders toward mutual understanding and continued success.

As W. Edwards Deming said, "Management of a system ... requires knowledge of the interrelationships between all the sub-processes within the system and of everybody that works in it." Establishing processes for decision making and control is essentially what is meant by the term "governance"; so stakeholder alignment in an IPA, PHO, MSO, ACO, or other entity involving professionals and clinicians, facilities, and administrators is closely related to organizational governance. Ideally, the mission, vision, and value proposition also align to focus everyone on what must happen for your organization to succeed. The ability to own and use primary source data and call upon accurate, current business information, when coupled with leadership skills of vision, communication, mission, and explain organizational strategy that maximizes the organizations' unique selling proposition (USP), which in turn enables the launch and manage a successful healthcare provider organization. Therefore, alignment is important because one cannot have an effective system without it.

Alignment is a key step toward the ultimate goal of integration. A provider organization can form an entity that has legal integration, clinical integration, or economic integration, but that is just the shell. It is easy to crack without having the alignment, because without true alignment, the shell is essentially superficial.

Effective integration between physicians or healthcare professionals goes beyond alignment and is achieved when the individual components of a performance management system operate as a fully interconnected unit. The success of this integrated operation comes through the application of long-established principles and practices in organizational development.

Physician–hospital integration refers to a variety of collaborative endeavors between the hospital and physicians, which may include, but are not limited to (Figure 1.1)

- Payment for medical and administrative functions
- Business practice alignment
- Medical directorships
- Recruitment assistance
- Professional services contracts
- IT support

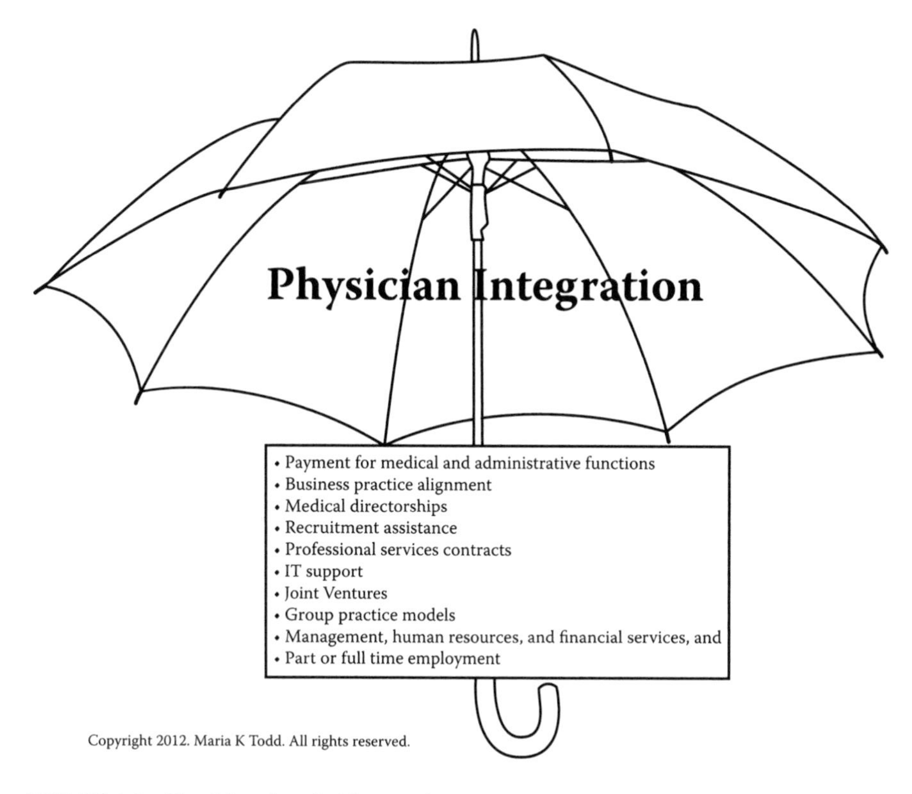

FIGURE 1.1 Physician–hospital integration.

- Joint ventures
- Group practice models
- Management, human resources, and financial services
- Part- or full-time employment

Wow! That's pretty much the whole universe of health delivery, is it not? So, to change any of these elements of the integrated health delivery "system" is a form of healthcare reform, right? Not just payment reform, but also operational reform. To accomplish this change in corporate culture requires sound practices and principles of organizational development and change management.

This system has several basic qualities:

1. The system must be designed to accomplish an objective, which is probably very closely related to its unique selling proposition.
2. The elements of the system must have an established arrangement.
3. Interrelationships ("alignment") must exist among the individual elements of a system, and these interrelationships must be synergistic in nature (WIIFM).
4. The basic ingredients of the process (healthcare delivery), including the flows of information, staff, materials, compensation, etc., are actually more vital than the basic elements of the system.

5. The organization's objectives are more important than the objectives of its elements, and thus there is a de-emphasis of the parochial objectives of the elements of the system.

So, in a nutshell, the integrated health delivery system takes such inputs as physicians, facilities, managers of the group, start-up capital, and energy and converts them into the products or services desired by consumers. The outputs include new research, well patients, improved medicine, trained doctors and nurses, cost avoidance, cost containment, inhibition of disease progression, continuity of care, lower prices, management and purchasing efficiencies, and other medical and wellness benefits. Okay, here is the sarcasm alert. But so what? Don't all these groups aim to produce these outputs? So, what is the USP of the provider organization you are contemplating?

Because health systems do not operate in a vacuum, they are described as open systems because they influence and are influenced by their environments through a process of interdependency, which results in dynamic equilibrium. Because the system achieves a steady state of continual interaction with its environment, it exemplifies dynamic equilibrium and cannot continue to survive without the continuous influence of transformational outflow.

This is actually more important than many people think because, if the organization does not offer this activity, the physicians who crave challenge, diversity, curiosity, and the ability to influence change in health status would disengage and leave out the provider organization in search of another one that meets their WIIFM objective out of sheer boredom!

The following characteristics occur in most integrated health delivery systems, regardless of what you call them or who the stakeholders may be:

1. Interdependence
2. Holism (the system is a whole, not merely the sum of its parts)
3. Inputs and outputs
4. Goal seeking: interaction between the stakeholders has to meet a goal (WIIFM)
5. Entropy: to keep the system operating, new energy and resources must be infused regularly
6. Steady state: the organization will be adaptive and self-regulating to maintain the dynamic equilibrium
7. Feedback of information regarding performance
8. Hierarchy and leadership
9. Differentiation (striving to move toward increasing uniqueness of the product to sustain competitive edge)
10. Equifinality (the result of several possible final states from an initial state, and a similar final state may be achieved from many different initial states)

This is not to suggest, as happened in the 1990s, that because all ten characteristics occur, that a group should hire a consultant who has a "one-size-fits-all" solution. However, out of both ignorance of how to select consultants and the ignorance and incompetence of many consultants that flooded the market in the 1990s, that was a

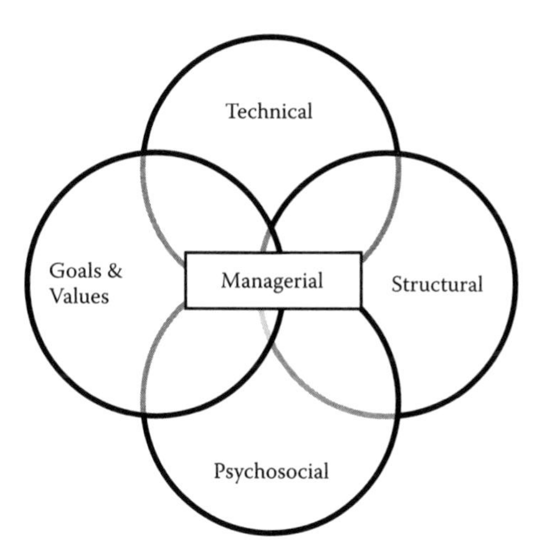

FIGURE 1.2 A sociotechnical set of subsystems for a globally integrated health delivery system®.

significant contribution to the demise of many IPAs, PHOs, and MSOs in the last healthcare reform era. The consultants you will need are like the architects, engineers, designers, and other professionals who have the ability to listen, and the experience to draw out what you want and to guide the process to what the members of the organization feel they want, to create the alignment and integration that "fits" their goals and objectives.

The way I process integrated health system development and operations, I view the organization as one that is an open sociotechnical system of coordinated human and technical activities. It does not matter if it is a hospital that employs physicians in Gurgaon, India or Greenville, South Carolina. It does not matter what you call it either. What matters is that it works.

Changes in any one subsystem of the organization will ripple throughout the organization, because all the subsystems are interrelated. In order to manage change or as we can call it "reform" of healthcare delivery, as the CEO of a globally integrated health delivery system®, a term of art registered with the U.S. Patent and Trademark Office (USPTO), I had to describe the five unique component subsystems that would define this term of art, as shown in Figure 1.2:

1. The *structural subsystem,* which includes formal design, policies, procedures, and so forth. This meant designing the organizational chart and describing divisions of work and patterns of authority.
2. The *technical subsystem,* which includes primary functions, activities, and operations, including the techniques, equipment, and so on, used to produce the output of the system. This is where the majority of medical tourism facilitators fall off the competitive field. They are, for the most part, completely unaware that these subsystems exist because they are unfamiliar with them in their entirety. It is also why I am "amused" when I hear of a trade association's efforts to create a certification program for these "facilitators" when the leadership of the association is itself seemingly unaware of these subsystems necessary to integrate health delivery.

3. The *psychosocial subsystem*, which includes the network of social relationships and referral patterns between physicians, administrators, allied health practitioners, and administrators, together with credentials vetting processes, privileging (roles), consultation reports, patient records, documentation, communication of best practices, and outcomes.
4. The *goals subsystem*, which includes the basic mission of the organization, including profits, growth, recognition, and business viability.
5. The *managerial subsystem*, which spans the entire organization by directing, organizing, and coordinating all activities toward the basic mission. The managerial function is important in integrating the activities of the other subsystems.

Regardless of which form or size of provider organization is developed, these component subsystems must be present. Otherwise, the system is likely not viable. It is the reason we author business plans, to act as a checklist and inventory for system development and product description, and it must be written by the stakeholders and managers who will have to live by it, live with it, manage it, and adapt it as the dynamic forces of change are applied to it over the life of the organization.

Over the next few years, and an impending election and acrimonious political agendas, we are facing the potential of unprecedented change in U.S. healthcare. If you look at that model in Figure 1.2, you have to add another larger sphere outside the Venn diagram, that of the *environmental suprasystem.*

People often asked me what will happen to my model of a globally integrated health delivery system® if the *Patient Protection and Affordable Care Act* (*PPACA*)* was repealed. Regardless of the Supreme Court ruling in favor of the PPACA, or attempts to repeal it, my model will survive. I am not as confident about ACOs and other provider organizations that are being designed to meet the needs of only one customer, be it CMS or the local HMO or PPO up the street, or a single employer in a town. When custom design work is endeavored, common business sense dictates that he who orders the bespoke solution should underwrite the entire cost of its development. Otherwise, that product is being built on speculation. What happens if that business prospect evaporates, abandons, chooses another product, or is deemed insolvent? What's left? That is exactly what people are asking me. They ask this not to be nosy; I am sure of that. They ask not out of concern for me; they couldn't care less. Most likely, they ask out of curiosity because they have not perceived the level of change that is about to unfold before them and the potential rapidity or how potentially radical the change will be. They still perceive the healthcare world as what once was. HMOs, PPOs, DRGs, CPTs, ICDs, doctors, patients, clinics, hospitals, and other indications of business as usual. If I sit and discuss it with them over a cup of coffee, the history is recounted and the epiphany begins. I will take you through it briefly in the next chapter.

* United States federal statute signed into law by President Barack Obama on March 23, 2010.

2 Independent Practice Associations (IPAs)

An IPA consists of a network of physicians who agree to participate in an association to contract with health maintenance organizations (HMOs) and other managed care plans. Although physicians maintain ownership of their practices and administer their own offices, the IPA serves as a corporate structure for negotiating and administering HMO contracts for its physician members.

In 1996, I wrote, "The time has come for medical providers to think of themselves more in terms of entrepreneurs in search of a place within a continuum of care, niche. Every other industry has turned to niche-marketing … why not private healthcare providers of all types?" In niches there are riches, right? Meh. Saying it doesn't make it so. The secret to success in the IPA niche is to have and put to use the trefoil of physician/member alignment, access to communication, and possession of data (Figure 2.1).

I also said that "The era of managed care is well underway and is not going to go away." I was right. It didn't. Instead, it reformed … again. Managed care continues to change and before we know it, we will be right back where we were in 1995 before I wrote the first edition of the book. But it continues to change at unprecedented speed, right? Absolutely! But for different reasons. And the most compelling reasons for this are associated with healthcare information technology, not smarter doctors, robotics, or wireless device technology platforms.

What was commonplace in 1997 is long gone, and new trends have replaced it. Thus, we need to begin to work with the system as it is and where it is going, as accurately as we can predict to improve care decisions and raise the bar on evidence-based medicine as we know it. This way, everyone benefits: payers, employers, employees, consumers, providers, and vendors who are interrelated and interdependent on a working and integrated system.

In response to pressures on the practice of medicine, new practice management styles and organizations continue to be created to meet market demands. Managed-care practices have encouraged the development of tightly integrated networks of physicians that in the past were only possible with closed-panel HMOs and other corporate structures to provide care for patients. In the past fifteen years, we have witnessed managed-care backlash from patients, providers, and even employers and insurers. Now, the buzzwords include "patient engagement," which in essence, really means flipping the responsibility for care rationing and decision making onto the consumer who complained about having it done for them. Early resistance of physicians to joining in administrative arrangements melted, the organizations rose up, were later abandoned, and are now seeing either redevelopment, reformation,

FIGURE 2.1 IPA success trefoil.

reorganization, or reanimation. Providers are once again adopting the philosophy of joining together and combining resources for survival and to improve market leverage, operating efficiencies, and economies of scale.

New business models are evolving, networks are forming as subgroups of previous loose affiliations, and other provider organizations, in an effort to form the next great thing. All the groups are integrating in order to find a comfortable way to align interests and cope with the eminent reforms on the horizon, the likes of which they have never witnessed in the past. Some are reactively traipsing into areas of uncharted jungles of administrative complexity; others, perhaps not moving fast enough depending on perspective.

In 1996, when I wrote the first edition, the Clinton Healthcare Reform Package was like a train in the distance (or the light at the other end of the tunnel?), and everyone in the tunnel (in the dark) seeing the light, simply wanting to take cover, and take steps to survive its passage without being able to see where they were going. Now it is Obama, PPACA, and new healthcare reform. It's kinda sorta like a revolving door, isn't it?

In the 1990s, United States healthcare providers looked at what was coming in the distance and moved like never before. Many got the idea to circle the wagons as providers and start building shelters that they had only heard about—some out of fear, some out of defiance, others out of ignorance, and a few with some well-planned forethought and market sensibilities. Some took the time to become educated about integration and capitation, while others bought assistance through hired consultants and lawyers to help them build a model that the consultants or attorneys had seen elsewhere. Many groups engaged their consultants with the responsibility to just build something—anything—but build it quickly. They scrambled to build a product for impending "managed competition." Remember that term? Were you there then? I was, and now I feel "old." I should tweet that term and ask how many people remember it; or post it in the Managed Care Contracting Group in LinkedIn. Wow! How did we live without artificial socialization like Twitter® and LinkedIn® and blogging? How did we ever communicate with one another?

"Managed competition" was a phrase we used widely during 1993, referring to a system proposed by the Jackson Hole Group (look it up in Wikipedia; it was not a rock band) that suggested that the individual employees receive a fixed sum from their employer and the individual employee chooses the health plan they prefer. Now we just call them HIEs, for health insurance exchanges. If the plan they chose cost more money than their employer's allowance, they paid the difference. The employee would have the tax incentive to select lower-priced options because they would only be able to deduct the amount of the lowest-cost option. The proposal's proponents believed that this would encourage individual consumers of healthcare coverage to be more price conscious (now we call that "engaged"), and they also believed that this activity would cause healthcare insurers to hold down prices of their plans to make them more attractive and, hence, more competitive in the marketplace (Figure 2.2). Because insurance under this proposed system was not tied to the employer,

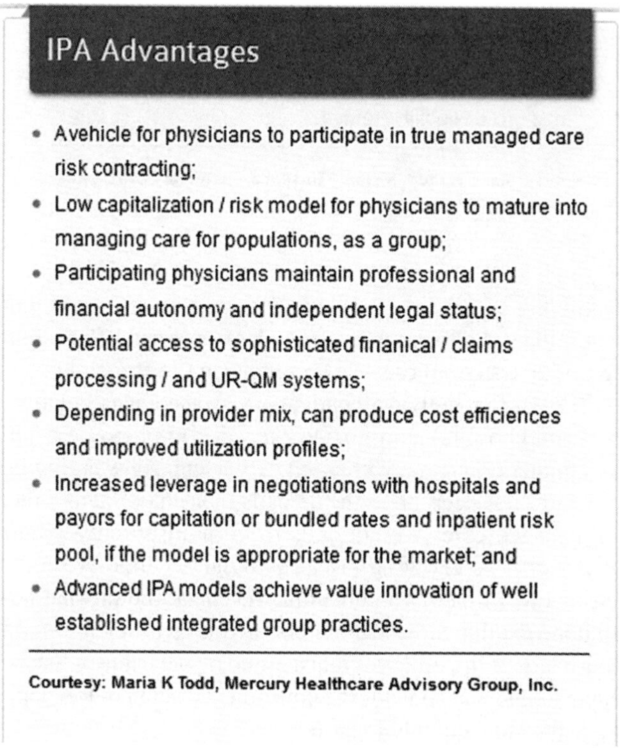

IPA Advantages

- Avehicle for physicians to participate in true managed care risk contracting;
- Low capitalization / risk model for physicians to mature into managing care for populations, as a group;
- Participating physicians maintain professional and financial autonomy and independent legal status;
- Potential access to sophisticated finanical / claims processing / and UR-QM systems;
- Depending in provider mix, can produce cost efficiences and improved utilization profiles;
- Increased leverage in negotiations with hospitals and payors for capitation or bundled rates and inpatient risk pool, if the model is appropriate for the market; and
- Advanced IPA models achieve value innovation of well established integrated group practices.

Courtesy: Maria K Todd, Mercury Healthcare Advisory Group, Inc.

FIGURE 2.2 Benefits of IPA.

IPA Disadvantages

- May be difficult to create same efficiencies as integrated group practice
- Difficulties in capital formation; weak systems infrastructure;
- Often unwieldy, specialist-dominated governance;
- Does not affect the expense side of physician practice legal issues;
- Antitrust: may negotiate capitation rates; same antitrust concerns regarding PPO-style discounted fee for service contracting;
- Legal form / corporate practice of medicine: because providing services, must comply with state law regarding form; generally organized as a professional corporation or medical partnership;
- Care must be taken to avoid aggregation and disqualification of participating physicians' pension plans; and
- Peer review and credentialing issues; potential for new malpractice liability created.

Courtesy: Maria K Todd, Mercury Healthcare Advisory Group, Inc.

FIGURE 2.3 Challenge of IPAs.

employees would have had portability (not lose coverage) if they changed jobs. Under this proposed system, there was no provision to set premiums that appropriately covered the risk of an individual patient or risk-specific population (Figure 2.3).

The physician and hospital administrator reaction to proposed healthcare reform with managed competition came with a jolt. My telephone was ringing nonstop for assistance; we could not keep up with the calls, nor could many other consultants involved with managed care. We got calls from multispecialty groups, solo practitioners, and others. The cry was, "Help us build a Clinic Without Walls!" The first lesson learned was to never assume that the client knows what he or she wants until you have done the due diligence of defining the terminology! All I knew about Clinics Without Walls at the time was that a group by the name of Sacramento-Sierra Medical Group had one and some of the constituents were not too happy at the time for whatever reason. How quickly I learned.

Once I clarified what kind of a "thing" they wanted to build (Figure 2.4), I got pretty much the same answers: "We want to build (1) something that allows us to

What Physicians Want in an IPA...

1. Preservation of clinical autonomy
2. Maintain business independence
3. Affiliation relationship
4. Negotiation leverage
5. Enjoy cost efficiencies
6. Preserve or grow market share
7. Enhance camaraderie between the providers
8. Maximize economies of scale
9. Access to better advisors than we can afford as individuals
10. Primary data ownership to measure and publish outcomes

FIGURE 2.4 Important points physicians need in an IPA.

preserve our clinical autonomy, (2) not become one corporation, and (3) have "kind of a loose affiliation with one another" that (4) can bargain with the payors with some clout. We also want it to be able to (5) facilitate sharing of overhead, (6) preserve market share, (7) enhance camaraderie between the providers, (8) take advantage of some economies of scale, (9) purchase contracting expertise that we might not be able to afford as individuals, and (10) have data to measure and publish outcomes." "Okay," says the consultant. "We can do that but it might not be called a Clinic Without Walls, if that is alright with you. How about if we call it an Independent Practice Association, or IPA?" This was usually met with, "Call it whatever you want, but can you help us build it? How much will it cost? And how long will it take?" responded the client. Nothing has changed except the person on the other end of the phone and what they want to call it.

Since then, the above conversation has been repeated nationwide. I responded to and was awarded more than 150 projects over the course of several years. To my amazement, as I reach out to several of the now-retired physicians on Facebook®, many of the groups I built then are still functioning and profiting, while others who wanted to build something against my better judgment (the customer is always right, right?) failed and, in some cases, went bankrupt for a number of reasons, the most frequent of which was lack of adequate capitalization, physician leadership, or because they built a thing that the market neither wanted, nor needed. Amazingly, fifteen years later, the situation remains essentially unchanged.

For example, one such encounter with a group in mid-1993 went like this: After meeting with a small group identified as the ringleaders or, in more professional terms, the "steering committee," and questioning them as to their goals and objectives, we began the actual development of the IPA. Together with their selected corporate attorney, we accomplished some of the initial objectives of starting off the group with its Articles of Incorporation (back then we formed S Corps, now they would be Limited Liability Corporations (LLCs) and require Articles of Organization), a legal document normally filed with a state agency that gives the entity birth in the eyes of the law. It basically declares who the parents are, how much influence the parents are legally entitled to, and several other declarations about the entity for everyone to see. The document is then accessible to anyone through the Freedom of Information Act (FOIA).

Next, we ratified the bylaws of the organization to be able to form and empower some committees so that the organization could do some business. This involved designing tasks to be done by committees and forming decision trees. I also assisted with the determination of governance structures. The attorneys provided the group with template documents that were then customized to the needs of the organization.

Once those preliminary steps were completed, the fork in the road was encountered. The direction the group took and still takes today is largely dependent on the level of commitment of the group to its leadership, to raising capital, and who is at the steering wheel of the project … and most of all, what problem it solves or unique selling proposition it brings to the marketplace. If the group wanted to move ahead with contracting with payors, they would generally design or have us design for them, a credentialing application and a provider services agreement. Most often, the group would then engage in some sort of social credentialing, by asking if everyone on the board is in agreement that the prospective member should be allowed membership in the organization. Unless there were any opposing votes or questionable concerns on the application, the applicant was made a member and granted privileges to see patients under the contracts that the group negotiated.

I should pause here because this too has changed over the years, but most physicians are unaware of the implications of the changes. In the old days (1990s), many IPAs were created in just the above process. Now, medical staff credentialing is an entirely different kettle of fish. If an IPA, MSO (Management Services Organization), ACO (Accountable Care Organization), or PHO (Physician Hospital Organization) wishes to vet the credentials of a provider, a process known as primary source verification is expected as industry standard.

For anyone in the United States, there is no doubt that the National (meaning United States for readers of this book from outside North America) Committee for Quality Assurance (NCQA) sets the standards for credentialing in U.S. managed care organizations. Funny, how if I read the standards for managed care in Malaysia or Australia, or South Africa, they are essentially the same. Any search for information on the subject directs the researcher to the expectations set by this formative organization. Defined as the "process by which a managed care organization authorizes, contracts with, or employs practitioners who are licensed to practice independently to provide services to its members," credentialing simply means making sure that a practitioner is qualified to render care to patients. Each managed care organization or integrated health delivery system (IDS) (IPA, PHO, MSO, PPO, ACO, HMO, etc.) sets its own qualifications and then structures its processes to ensure that the practitioners meet these qualifications. It also assumes liability for the integrity of the process and any errors or omissions as they relate to the safety to patients represented by the organization in offering the panel as a "product" to the marketplace. The topic of credentialing is presented in deeper context later in this book.

A typical IPA provider services agreement and credentialing application can be found in the appendices section at the end of this book. For the most part, many IPAs had intentions to follow through with their goals and objectives beyond that point, but many never really made it.

Once they had an organization, many went to the negotiation table with the payor and never finalized their quality improvement documents, utilization plans, credentialing policies and procedures, etc. They merely existed as localized groups that had been gregariously formed, open to all licensed, interested physicians, whose goal was to retain autonomy yet have strength in numbers for bargaining power with managed-care contractors—like a symphony orchestra full of soloists! Now, payors, having accepted this defense tactic, want a more complete package, including cost-effective players, with emphasis placed on financial and quality outcomes.

SECOND-GENERATION IPAs

Next we began to see the proliferation of the second-generation IPA, one that meets this criteria and possibly (probably) engages in capitation. These groups are usually a smaller, more quality-oriented panel of providers, willing to police their own members—a "chamber orchestra" willing to be selected among the best players on an audition basis.

As healthcare vendors, you must anticipate the needs of your payer (insurers, TPAs [third-party administrators], and employers and labor unions) and consumer audiences, and keep your minds open to the possibility and likely reality that these provider-based organizations will form and grow into competitive groups, each with their own niche focus and service style—their unique selling proposition, or "USP."

Successful implementation of these provider organizations in the small community requires the essentials of availability to patients, consistency of practice, and a friendly hometown approach. (Now they call this concept a "medical home"—hmm, how quaint. I can't begin to imagine what it will be called when I edit the third edition!) The group must have substantial clout with the hospital, specialists, and insuring or paying organizations, substantial sharing of risk, overhead, and the capital to handle the expense of management skills for the complexities of the business of medicine, which are an essential driving force. The ability of primary and specialty physicians to work together effectively requires a vision of the future and the administrative leadership of the physicians and the hospital administration, with separate, yet balanced power. Remember that we manage things and lead people!

In many cases, the IPAs in the local neighborhood will grow and partner with other healthcare providers such as hospitals and management services organizations.

A new organization, whether PHO, MSO, or other set of letters, must be given birth by some parent, and a local IPA is as good a start as any. However, the group must be well designed with a more homogenous mix, made up of providers with different interests, locations, hospital affiliations (where applicable), and specialties. The common vision to organize an effective single business entity to maximize economic clout, provide quality care, and yet preserve some of the traditional values of autonomy is necessary. Without this "bigger picture," actions fall short of everyone's need. If you cannot satisfy a need, there is no reason for product development. The marketplace has no need for another brand of widget!

Second-generation IPAs have historically been loosely organized, overly dependent on volunteer time, and fragmented by organizing around hospital affiliations or particular insurance plans. Usually, no single IPA can offer everything to align with your preferences and needs as an individual healthcare provider. You may find yourselves required to join multiple organizations to achieve your goals. While counterproductive, it is reality. In essence, you allow the market to choose from you and yourself. You will compete with yourself, and the payor who wants you simply picks networks because they access your services at one price or another simply by picking either IPA or the local PHO.

For example, organizations meant to provide advice on contracts could only provide broad concepts rather than getting into the nitty-gritty of negotiation or strong advocacy because of anti-trust concerns that severely weaken any collective process or structure, especially in IPAs.

The potential of a new IPA can be realized by developing an innovative product for the market. For example, an organization of healthcare providers from different parts of the metro area, who work with different hospital staffs, and gives the group protection from various shifts and changes of contracts from one hospital to another. In this way the group can always provide care, no matter where in the metro area or what hospital facility. Employers must no longer be concerned about finding quality physicians in various parts of the metro area.

The group must produce value. Now, don't get all egotistical! Think like an entrepreneur! More specifically, the simultaneous pursuit of differentiation and lower cost. Your IPA will realize high growth and profits by creating new demand in an uncontested market space, rather than by competing head-to-head with other IPAs for known customers in an existing market space. This is shown in Figure 2.5.

Step 1

The first step in forming this new product (regardless of what you call it: an IPA, an IDS, an PHO, a second-generation IPA, a super PPO, etc.) is to abandon the idea that "Doctors do not agree or get together on anything." Perhaps the doctors who do not agree are not mentally ready to change enough to survive in the present and future market-driven and legislated reform. There is no need to convert them; simply move around them. Nowhere is the phrase, "Lead, follow, or get the hell out of the way!" more appropriate. You cannot help those who will not help themselves.

Creating Value Innovation in an IPA

Eliminate	Raise
Which factors can you eliminate that other IPAs have long competed on? **List those here:**	Which factors should be raised by your IPA well above industry standards? **List those here:**
Reduce	**Create**
Which factors should be reduced by your IPA well below industry standards? **List those here:**	Which factors should be created or offered by your IPA that the industry has never before offered? **List those here:**

Courtesy: Mercury Healthcare Advisory Group, Inc.

FIGURE 2.5 Value innovation in an IPA.

Step 2

The next step is to grow politically. Economics will follow the politics. Pick leaders who will bring small groups of physicians with them. Gather the ringleaders and arrive at a consensus on the vision, reasons, and needs for the development of the group structure. Never underestimate the power of politically active, leadership-oriented individuals helping to drive the process.

Step 3

Third, do not wait! Take individuals who are ready and go ahead. Leave the door open for those who want to come later. Do not assume that there is a magic number of individuals required to start the group, or a logical set of specialties. Chances are that if you build something appealing to the marketplace, the providers will bang down the door to get in. Perhaps more providers than you might need or are able to support.

Step 4

Fourth, select colleagues based on the general quality of their practice. Do not prejudge; if the practitioners are basically good and committed to the process, invite them in. There is no requirement to take everyone; but on the other hand, do not allow personality differences, personal perceptions, or biases to interfere with inviting someone into the group, if only on a provisional basis.

FINANCING AND MANAGING THE IPA

Use an appropriate set of screening mechanisms for new members to join the group. Obtain all the rudimentary information that is required to join a typical managed-care payor organization. Try to ensure that the members are at least currently economically viable. The reality of withholds and administrative costs during transition must not be financially overburdening.

Maintain existing working relationships and referrals. Do not require referrals within the group as an organizational issue. It may become a problem with Medicare, Medicaid, and other self-referral legislative issues already in place. It comes naturally that the members will want to support their partners, particularly as they develop new practice patterns. Do not damage existing relationships by exclusion. Just keep quality and cost-effectiveness front of mind when reviewing out-of-network referrals. Be quick to point out recruiting opportunities when outside referral work is well done and cooperative in nature. It bears repeating that one must remember to plan to remain in conformity with the Medicare/Medicaid anti-kickback restrictions that prohibit remuneration in cash or "in kind" for the value of the referral stream when negotiating with ancillary and hospital affiliates. The OIG* Fraud Alert titled "Hospital Incentives to Physicians" will help apprise you of the necessary issues for consideration. This issue is addressed at length in its own chapter ahead.

Construct the business so as to maintain a reasonable degree of management autonomy for the physicians over their immediate staff and facility in which they work.

As far as money management of the basic organization, pay the IPA bills first! No process is more motivating than to realize that the bills get paid first and the members get paid second. It creates a strong, financially sound group with good relationships with the creditors and businesses with whom the group works. Usually, the IPA does not pay its members for anything unless it bills for services under the IPA tax identification number. This is an activity usually reserved for later in the development stage when an MSO is formed or employed.

Do not mess with billing. Unlike the merging of existing practices, whereas a new corporation is created and the new corporation would require new provider numbers, new IRS numbers, and even perhaps Medicare numbers by site or facility, depending on the intricacy of the infrastructure, this is not so in the typical IPA. In an IPA, the practices do not merge, but instead retain their corporate autonomy. Therefore, in most cases, joining an IPA does not require that an IPA member start changing numbers or that one must dismantle an existing billing format within an individual office.

However, there is a need to conform to some centralized or uniform billing practice, as well as a uniform fee schedule. Be advised, however, that the fee schedule part cannot be done without the proper economic integration and then only with caveats of antitrust laws as applied to the healthcare area. In most cases, the IPA is kept "poor" and maintains "shallow pockets" for the purpose of liability risk management. In this case, it is necessary to keep the economic integration at arm's length by designing an MSO that better meets the objectives of economic integration and thereby addressing pricing, negotiating, and contracting issues at that level. I discuss that later in this book.

Also, keep in mind that a payor may offer a fee schedule that everyone in the IPA may agree upon without meeting the standards of economic integration. However, this does not imply that the IPA has negotiation "clout." It is still not possible for the IPA to propose a set price to the payor so that the payor can access services unless

* Office of the Inspector General.

anti-trust guidelines for economic integration have been tested and conformed with. Nor can the IPA collectively disagree with the payors' offers to prevent contracting problems of "naked price fixing" or "boycott." These issues come up when we see a concerted effort to restrict competition in the marketplace arising from the activities of the members or agents of members in an IPA without following the rules about price fixing and boycott, or monopolistic tendencies. Some groups have engaged both managed-care and legal counsels, who are not experienced or well-versed in antitrust, to negotiate for them, so as to let the consultant engage in an act of (albeit cerebral) sharing sensitive information and making suggestions for a group to follow. They might all be playing a few rounds of golf at "Club Fed"!

Do not assume that an IPA needs to create a centralized billing structure overnight. The realization of the need to deal with this issue usually comes about the third month of development. At that point, it is no longer a sales job, but a question of how to achieve co-development of an IPA and an MSO. I discuss more billing issues in the MSO section later on in this book.

PICKING YOUR CONSULTANTS

Choose a law firm that is familiar with healthcare but that can provide a variety of services, minimize the learning curve on healthcare issues. This is not a generic business entity similar to a lawnmower sales center. We not only have normal contract law to deal with in this business, but we have so many regulatory overlays that a specialist is needed with expertise developed in this area. In my travels, I have come upon too many generic lawyers who find that a start-up entity is a good opportunity to learn the healthcare area as they go, yet they charge the client on an hourly basis while they spend their time researching and reading what an experienced health law expert already knows. A start-up IPA needs the best expertise its money can buy in the areas of consulting and legal assistance. You either pay in sweat equity and cash, or cash and sweat equity!

Make sure that you perform all the due diligence in requesting references, published works in the subject area, and talk to former clients and colleagues. Make sure that you are not engaging a consultant or attorney who only has one solution, one style of project, and one fixed rigid way of seeing the task at hand. Each project is going to have similarities to others but each one is going to have its own design specifications because of differences in participants' cultures, communitized needs, start-up and operating budgets, and educational orientation.

OPERATIONS MANAGEMENT FOR THE IPA

View the corporate office as a service center, and put the patient at the top of the process. If a corporate structure regards itself as the center of authority, as opposed to the center of service, it is not focused on its primary responsibility. The usual task of utilization monitoring and management, as well as outcomes, patient satisfaction monitoring, and measuring, referral management within the utilization management guidelines, and day-to-day operations management is best suited for the MSO. However, as an IPA, it will be necessary to develop those items to be measured, monitored, marketed, contracted, etc. Then the MSO has something to do!

Administrative Staffing for the IPA

Most IPAs do not need a large support staff to operate. Many start-ups have operated with a part-time administrative assistant, working as an independent contractor. This person may serve as telephone liaison to the payors, secretarial support to the officers and directors, and may take care of handling IPA mail as it arrives. Once the organization gets larger, a full-time (or part-time and full-time) person may be needed to assist with credentialing, mailing and faxing of contracts, etc. It may also hire an executive director who answers to the board and carries out the day-to-day business affairs of the organization. The IPA should probably avoid clinical personnel who make clinical decisions or lay hands on patients because of liability concerns. When hiring an executive director or an administrator, do not look only to a candidate with previous experience in contracting—this is the error most often made. Find a candidate with group practice exposure, one who is current, innovative, and has a variety of operations management experience—a leader and motivator with a "big-picture" mentality. Find support staff to carry out the myopic details under direct but minimal supervision.

Governance Issues for the IPA

Governance of the organization is critical. Resist the temptation to slot too many positions and to guarantee the representation of specific interests. Focus instead on the individuals who are elected, and expect them to represent the entire group structure and culture. Give the board substantial power so that it can act aggressively on everyone's behalf, yet keep them on a short leash to allow adequate opportunities to replace individuals as necessary.

It does not help an IPA for the members or shareholders to be making every decision; it interferes with efficiency and does nothing to gain the advantages desired in the marketplace.

Start-Up Capital

It is going to take time, effort, and seed money to start up an IPA. Be cautious about seeking help from hospitals and other groups that might provide additional early cash. From experience, I have seen groups develop internal cohesiveness if they invest their own time, sweat, and money in the organization. In Figure 2.6 I provide some 2012 estimates for a high-level budget for start-up IPAs. As market and geography and technology change, the numbers will also change. As this book is intended for a life span of several years, do not forget to adjust costs for inflation, as needed. Please remember that the budget is differentiated from other start-up types of integrated delivery systems with other goals and objectives.

It is necessary to look at Figure 2.6 and then rehash what the organization really wants to accomplish in its first year. As a consultant, I would suggest staging the computer system for a later time, and working on one to three terminals for the first year to be able to do word processing, dues, accounting, etc. I would defer on lots of

IPA Proforma Budget - Startup

Budget Items	Minimum Expense
Legal Expense - filing fees	$2500
Legal Expense - counsel	
Consultant Expense - organizational development	$100,000
Accountant Expense - counsel	$10,000
Actuarial Expense - if capitated	$30,000
Commercial artist - logo development	$250
Credentialing & Privileging Expense - software	$15,000
Credentialing & Privileging Expense - vetting fees	$75 per doctor
Wages, commissions (staff, brokers)	
Telephone	
Postage	$500
Rent & Office Expense	$20,000
Office Equipment & Supplies	
Printing	$5000
Officer & Director Insurance	$2500
Errors & Omission Insurance	$2500
Reinsurance (if capitation deals are negotiated)	
Computer Hardware & Software	$6,000
Website Development & Hosting, Cloud Space	$30,000
Administrative staffing	
Miscellaneous expense	
Total:	

Courtesy: Maria K Todd, Mercury Healthcare Advisory Group, Inc.

FIGURE 2.6 IPA budget startup.

staff and choose to use a part-time independent contractor secretary, remote voice mail, and e-mail, from the telephone company or other vendor e-mail, thus saving lots on postage, office equipment, paper and printing costs, etc. I would also suggest a deferral on errors and omissions coverage until you need it. This is very different from the officers and directors coverage you do want to purchase once you elect the board.

This would make the first-year budget pro forma to less than half of the original quote, more feasible, more palatable, and achievable. The way I laid out the pro forma in Figure 2.6, I guesstimated on several items from experience for the first-year cash projections. Not all of the cash is needed up-front. Some of it is needed to begin to develop the organization. Not all the big-ticket items are paid in full upon signing a contract with the consultants. Pay in stages as deliverables are produced, but you will need to pay a retainer to get started and you will always need to pay for travel and expenses as they are incurred. Expect to be charged Business Class airfare for flights more than two hours, as many consultants with high-volume travel have issues with DVT these days, and it affects our workers' comp premiums, so we mitigate by taking better care of ourselves. Expect us to want to do more Web-based meetings and less travel to your location except at the beginning. Telephone bills need not be hefty these days, but you will pay a higher rate for commercial service than residential and wireless services. You may consider an advanced VOIP* system.

Solvency Standards

You may wish to establish solvency standards for the IPA to avoid the delusion that your organization is viable if it is not.

These standards should include requirements to

- Pay all downstream (i.e., to contracted practices) claims on time.
- Show that they are measuring incurred-but-not-reported claims accurately.
- Demonstrate positive net equity and sufficient working capital.

If a group or IPA cannot do all of these, it should not accept risk contracts and it should never do so without adequate reinsurance coverage.

In the 1990s, MedPartners, FPA, and risk-bearing IPAs and medical groups in California and most other states were not held to capitalization and solvency standards traditional in the industry. The bankruptcy of Genesis Physicians Practice Association, a 960-doctor IPA in Dallas—$7 million in the hole—and another in Denver—$8 million in the hole—might have been preceded by warning signs, if not prevented altogether, if they had adopted or been subjected to state regulation. Because groups are unregulated, data about their finances are not publicly available.

Now the next hurdle: Where do you get the money? Seed money is necessary at first because you will not be able to sell shares to anyone until the organization is at a stage where it can do so. The Offering Circular† or Private Placement Memorandum (PPM) development is costly to draft, have reviewed by specialized counsel, and filed.

Therefore, you must rely on the prospective membership who are not members until they have been credentialed and offered a contract to participate. This means that you may have to develop a means to raise capital.

One group I worked with in 1993 designed a structure so that $850 was initiation fees (nonrefundable), $200 was dues for the first year, and at least another $200 was put into an escrow account toward shares when they became available. The initiation fee was accompanied by a pledge of at least thirty hours of sweat equity so that the work was divided among all membership prospects and there was little apathy or blaming. This enabled a group of 140 physicians to raise $88,400 and hold lots of committee meetings to kick the organization off to a fine start. The additional $200 membership and equity cash was paid upon completion of credentialing and formal acceptance into the group. It also loaded the group with 420 hours of uncompensated developmental time from stakeholders with something to lose in case the apathy mind-set became a problem. The stakeholders were quickly allowed to choose

* Voice Over Internet Protocol.

† A legal document offering securities or mutual fund shares for sale, required by the Securities Act of 1933. It must explain the offer, including the terms, issuer, objectives (if mutual fund) or planned use of the money (if securities), historical financial statements, and other information that could help an individual decide whether the investment is appropriate for him or her.

which committee(s) they wanted to participate in, and we gave each committee a rundown of what there was to be done and in what time frame. The following committees were necessary to develop the organization. Ah, history! Now adjust those figures for inflation some twenty years forward. We estimated the value of their hours at $50 per hour. A handy tool is available to calculate for CPI Inflation online (see <http://146.142.4.24/cgi-bin/cpicalc.pl>).

Steering Committee

These individuals are the ringleaders who happen to work well together as a team. Hopefully, they have taken time to obtain some education about managed care, integration, contracting, and capitation; have charismatic personalities and good leadership skills; and can read marketplace dynamics and activity in the community with an almost innate sense. From this group the officers and directors will emerge.

Bylaws Committee

This committee is headed by at least one steering committee member who is supported by at least two other prospective members of the IPA to review for current legal and structural conditions. Their actions will set the tone for how the IPA conducts its business affairs and governance issues.

Termination with and without cause may be required, and there should be some published policy in the bylaws if the Provider Services Agreement mentions that such a policy exists. Items to consider include stock transfer, membership rebates on unused portions, return (if any) of initiation fees, retention of shares by non-members, etc. A good source of information about termination policies might be gleaned by the IPA from an informational package sold (and available to non-members as well) by the California Medical Association, in Sacramento, California.

Membership Committee

This group makes suggestions (nominations) for who will be invited to join and what the conditions of membership will be, and communicates this to the bylaws committee for inclusion. The membership committee also designs the Provider Services Agreement to be used as the contract between members and the IPA.

Utilization Management Committee

This committee will perform statistical functions that will set standards and eventually evaluate productivity by Current Procedural Terminology (CPT) code to determine the activities of the group as a whole. This should be headed by at least one board member and supplemented in the physician setting by a team of physicians including specialists, sub-specialists, and primary-care and hospital-based physicians. In an IPA of other types of providers (that is, physical therapists, occupational therapists, and the like), it should have a composite team made up of members with varied expertise. Understand that this committee does not do the day-to-day monitoring, but reviews and makes decisions once the MSC is capturing data and reporting to the board for review and action. This is why it is necessary to establish only a brief policy statement to begin the organization. So many IPAs mention that membership shall abide by the IPA's UM/UR* policies in their provider services agreements, yet many organizations that have been in business have them published in the event a diligent member might want to review such a document prior to agreeing to abide by it sight unseen.

The UR Utilization Management Committee will set forth standards and practice protocols or guidelines for more costly procedures and diagnoses to position for per-case charges and capitation rates that are inevitable. Among other things, under-/over-utilization and adverse selection monitoring, catastrophic case management, C-section rates, hospital utilization, specialty utilization, and emergency utilization must all be monitored by this committee.

Quality Assurance (QA) Committee

This committee will set forth and adopt quality management guidelines and standards, with the power to investigate quality concerns and issues and set forth corrective actions and requirements as may be necessary from time to time. This committee shall also set forth monitoring guidelines to prove implementation and compliance with quality measurement methodologies.

Quality assurance programs in IPAs present a distinct set of opportunities and concerns when compared to group and staff model HMOs. The delivery and reporting systems are complex and diverse. Many managed-care entities are now conforming or adopting as their standard of quality measurement the National Committee for Quality Assurance (NCQA) guidelines. The NCQA accredits HMOs only; however, many IPAs and PPOs are adopting the quality measurement guidelines to promote a "me-too" attitude. Each system has its own design and thus its own prime objective, with utilization management monitoring tied into the quality assurance program, and based on actuarial data and weighted according to capitation and budgeting philosophies.

Fundamental to any Quality Improvement/Quality Assurance (QI/QA) program is the screening of physicians who participate in the WA or any other form of alphabetic acronym involved in managed care. Criteria for acceptance beyond provisional status include having proper credentials, accessibility, and quality performance. A medical director, executive director, or administrator must also review the physical location where services will be rendered, the available equipment, and discuss with the provider and staff what will be required for a comfortable relationship with the group.

The HealthCare Standards Committee, usually a part of the Quality Improvement Committee or under its oversight as a subcommittee, performs standard setting (noncontroversial, measurable and auditable, importance, and potential to show improvement should be the objectives of such a program that this committee oversees). Quality assurance ratings should be determined by medical record audits, member surveys, member transfer rates, and managed care philosophy consensus.

There are many elements of the private practice of medicine that cannot be duplicated in any other setting and that, by their very nature, lead to a high level of physician and patient satisfaction. Intrusions, however necessary, into private physician

* Utilization management/utilization review.

practices should enhance rather than compromise these assets. Your QA programs should be constructed in a way that lets the physicians test themselves against a system that confirms the value of their individual practices and recognizes individual merit.

Finance Committee

The Finance committee sets forth a plan of action for money management, funding, and for the consideration of any nonstandard payment mechanisms and carve-outs from capitated arrangements and deviations from established CPT code values. The Finance Committee will probably also oversee purchasing and capital acquisition matters as well as banking relationships and investment strategies.

Credentialing Committee

This committee makes decisions on what the minimum standards of credentialing of the IPA will be, and hopefully (hint, hint!) keeps its decisions commensurate with both standards set forth for accreditation by The Joint Commission and the National Committee for Quality Assurance (NCQA). This committee either delegates the task of credentialing to one of the many credentialing houses nationwide, or oversees production of primary source verification as required by the above organizations from an in-house effort. The Credentialing Committee will also have the responsibility to design a new or adopt an existing credentialing application that will be used for all prospective members to complete prior to actually being accepted as members of the IPA.

As for what to request and verify, start with an application that covers the following:

- Name
- Address
- Telephone number
- Home address
- Home telephone number
- DEA registration
- Board certification status
- Malpractice insurance
- The yes/no questions in all managed-care agreements
- Proof of licensure
- Authorization to check references with impunity
- Request for peer references
- Hospital privileges
- Any gaps in employment over six years

which usually looks the same for most physicians. Significant additional data can be obtained by visiting the physician's office and by obtaining both written and verbal recommendations, the latter being more candid. A review of continuing medical education (CME) courses will hint at a commitment to quality care and current level of additional training beyond residency.

Other Concerns Relevant to Prequalification for Membership

It is necessary to review the prospective member's office policies and procedures, and to establish some basic criteria for membership.

Access Issues

Service locations should be open for a minimum of twenty hours and at least four days per week. After-hours coverage must be confirmed in writing with confidential home telephone numbers and other reach lines of all concerned listed in the cross-coverage plan. A review of who provides the after-hours coverage is also important. Non-members providing cross-coverage should be credentialed by the IPA as many managed-care contracts demand that the responsible party who is contracted warrant and guarantee the performance of anyone performing cross-coverage and backup coverage duties. A great deal can be determined about a practice by reviewing the appointment schedule.

A provider with long appointment waiting times or too many visits per hour may infer that he or she is not a good candidate for IPA participation. The NCQA has established performance standards with regard to waiting times around scheduled appointments. Most HMOs now try to establish similar waiting times as part of contractual performance. The most popular performance standards I have noticed in recent contracts are right around fifteen to twenty minutes.

Providers should have the capacity to enroll at least 500 new patients. The ability to ensure adequate membership in each provider's office is an important determinant of compliance with the plan's quality assurance plan.

Medical Records Review

The medical records review is significant in several ways. A credentialed Accredited Records Technician (ART) experienced in utilization review and QA should make the first-pass inspection, followed by the Medical Director. The review is one of the few times a physician will review the records of other providers in their own office; it symbolically indicates that the records of the providers will be reviewed in the future, and it is also an efficient way to determine if the records will pass muster under an audit of the organization's QA system.

Recertification and Recredentialing

The QA Committee must recertify offices on a periodic basis. Most do this every two years on a rotating basis so that the task is not so burdensome. In addition to adherence to the initial criteria, evaluations must cover several categories with an annual site visit being carried out before submitting a recertification recommendation to the QA Committee. The committee may decide to recertify, place on probation, or recommend termination

Grievance Policies

Grievances should be categorized as administrative and medical. Grievances should not exceed 2:1,000 members per year optimally. It is usually the Quality Assurance/

Quality Improvement Committee that develops the grievance policies, although if enough help is available, a separate committee might work a little faster.

Other Operational Issues

Member satisfaction and patient satisfaction are usually relegated to marketing committees, of which there may be up to four subcommittees: provider relations, patient relations, public relations, and hospital relations.

In conclusion, the above overview should help to hone the plan a little more as you begin the arduous task you are about to endeavor. There are many ways under the above topics in which assistance from a practice management specialist/consultant would enable you to provide clinical services while receiving help in the developmental stages of the operational planning of the IPA's activities and protocols. The learning curve on the administrative side would otherwise be very costly rather than retaining an experienced team of consultants. In later chapters I review some of the other activities necessary for all groups, whether IPA, PHO, or MSO.

3 Physician Hospital Organizations (PHOs)

A PHO is an organization that unites a specific hospital and certain physicians through a contractual relationship.

Every hospital continuously seeks to strengthen its physician loyalty bonds by forming something, usually Physician Hospital Organizations (PHOs). Increasing competition in the managed-care marketplace is forcing both providers and facilities to reevaluate their alliances and find ways to create a new product to take to market while complying with anti-trust restrictions. Now the buzzword for about 43 percent of hospitals is also ACOs, Accountable Care Organizations (see Chapter 4).

We should begin by first establishing what exactly is meant by the term "PHO." A PHO is an organization that unites a specific hospital and certain physicians through a contractual relationship. It is usually owned by both the hospital and the physicians, and in many cases throughout the United States is found to be a 501(c)3, not-for-profit organization. It is an entity that can negotiate with Managed Care Organizations (MCOs) and employers, while allowing physicians and hospitals to coordinate delivery of care through a jointly self-managed system, rather than allowing the payor to manage from the outside looking in.

PHOs are able to offer more enticement to MCOs and employer groups who prefer to contract with a single entity for a full package of health services rather than draw up and maintain several contracts with various providers individually. The enticement is in the reduction of labor-intensive management, utilization review, and quality assurance tasks. There are still a few managed-care payors that prefer to maintain what I call a "divide-and-conquer" philosophy in the belief that they can negotiate better (usually unilaterally for the payors) with individual providers and manage care better (for whom?) than most providers.

Careful consideration must go into the design and management of a PHO to provide premium quality of care and financial and management efficiency. Oftentimes, it may be inefficient for a small PHO to manage itself, as there may be goal conflict between the physicians and the hospital, resulting in confusion and anxiety among the participants in the group. Most often, these conflicts arise in the areas of financing, revenue allocation, administration style, subcontracts, carve-outs, and control.

Initial financing of a PHO is a dilemma. Hospitals usually have more ability to come up with start-up capital in much higher amounts than individual physicians. However, in order to have balanced control, the hospital should never be the sole source for financing. Both the physicians and the hospital should have equal risks and incentives, with negotiations keeping vigilant of that objective. Both groups should also share equally in profit and loss in order to conform to anti-trust matters.

PHO REVENUE ALLOCATION

Revenue allocation, otherwise known as "who gets what and how much," is another hot topic. MCOs like to make capitated/risk-sharing arrangements with PHOs. It makes for easier bookkeeping for the MCO, and makes for grizzly nightmares for the PHO that has not done its homework in cost accounting and actuarial forecasting. It is even worse for the hospital that has just recently launched a marketing campaign to become recognized as the community leader in cancer care, high-risk obstetrics, or cardiac specialty care, without also sending a message that the hospital provides great outpatient care. In order for capitation and risk-sharing to work, all the players on the team must be playing the same game with a common philosophy for utilization of resources, including specialty referrals, ancillary services, and technology.

When a PHO is able to attract from a payor, usually it is able to attract anywhere from 70 percent to 80 percent of premium dollars if it is really prepared to deal. By prepared to deal, I mean not only a hospital, but also physicians, ancillary providers such as home health, DME/HME*, respiratory care, pharmacy, lab, radiology, optometry, chiropractic and podiatric services, mental health services, ambulance and air rescue services, trauma and tertiary care arranged, etc. At that point, the budget is usually allocated at about 35 percent to hospital and other services and a professional services budget of about 28 percent to 32 percent. The other dollars are usually spent on several things such as reinsurance, plan management, referral management, etc.

PHO DIRECT CONTRACTING

For this reason, many full-fledged PHOs that have already created the MSO† component, either within or without the PHO structure, often seek direct contracts with local businesses. This has been seen more in the western United States than anywhere else. I have also seen a lot of this activity in Texas, where for some time the attorney general has been waiting and watching as PHOs engage in activities that might be construed as being in the "business of insurance." For the present time, I have noticed that the message from most insurance commissioners remains "no direct contracts that involve capitation on a direct basis." Why not?

The reason is simple. Say a PHO that does not go through the rigors of licensure as an insurance product might take capitation from the direct contract source. What happens if the money has been paid as capitation but for whatever reason the PHO cannot render services as necessary? Who does the payor turn to for remedy? Who pays for the covered services that employees, union members, or whoever needs them from within this direct contracted group uses them? This is especially important when contracting with Employee Retirement Income Security Act (ERISA) (self-funded employer) groups. These groups establish a 501(c)9 trust subject to U.S. Department of Labor guidelines, one of which is to maintain a certain threshold of liability per employee that would be transferred at the first dollar of capitation paid to

* Durable Medical Equipment/Home Medical Equipment.

† Management Services Organization.

the PHO. Therefore, ERISA plans, in most cases, may be prohibited from capitating a PHO directly without working through a licensed insurance company or HMO*.

They can, however, contract direct on a fee-for-service basis. They may even wish to use the MSO as their Administrative Services Organization (ASO), Third-Party Administrator (TPA), or Employee Benefits Administrator (EBA) if the services are sufficient, essentially the same, and priced right. Sometimes the user fees that ERISA plans pay to PPOs†, TPAs, and the like to access the discounts are upward of $28 to $60 per employee per month (maybe even more), depending on the site of the client. It all depends on the services being sold for the price.

It is possible, however, for any payor, ERISA or otherwise, to contract directly with a PHO on a fee-for-service basis without going through an established managed-care organization such as an HMO or PPO. We see this activity on the rise as many employer and purchaser coalitions become more and more popular throughout the country.

NEGOTIATION AND PROJECTION HINDRANCES

What hinders the negotiation process in the early days of a PHO for the physician and outpatient is the lack of qualified and verified data from such a variety of players. Because none of the physician's bill is one comparative entity as a whole, it is difficult for the hospital to assess referral patterns and practice styles from a purely statistical point of view. Without knowing peer comparisons adjusted for age, sex, catastrophic cases, and adverse selection, it is difficult to massage what little data are available and come up with a rationale for any cost-based system of revenue allocation for both the hospital and the physicians of the PHO.

Further differences can be demonstrated in a system with minimal internal medical management from those with a medical management system designed with a high degree of complexity and the relative discounting acceptable to participating providers. It is difficult to fashion a plan accounting for differences in revenue allocation between providers of similar type as well as types of providers. Everyone must work together to achieve consensus to the intention and methodologies used to calculate the allocation of revenue and then adjust—psychologically and professionally—any necessary personal changes to work the plan. Communication is key here, with all involved parties receiving and understanding documentation of the methodologies used to calculate the revenue allocation.

DIFFERENCES IN ADMINISTRATIVE STYLE AMONG MEMBERS

Administration style is a major difference because physicians generally are products of a top-down, uncomplicated (monarchical) management style very different from that of the integrative or management style needed to run a hospital. Both the physicians and the hospital must think along the lines of their historical adversaries, the insurers, as the PHO is essentially a smaller insurance company or MCO. It is often

* Health Maintenance Organization.

† Preferred Provider Organization.

noted that self-administered PHOs might not demonstrate the maximum efficiency if membership is small. Whether the PHO is self-administered through its own MSO or is administrated by an MCO, reporting mechanisms are crucial to regularity and timeliness, and a clear understanding of what is being communicated by the reports to all participants.

Often, the reports are fairly periodic (usually monthly) and less than timely (usually delayed by a full quarter) to evaluate services incurred but not reported (IBNR), but are generally so full of percentages, rankings, and gross number calculations that few recipients of these reports can make sense of what it is they are trying to interpret. The educational value of these reports, although well-intended, is many times tossed in a drawer for further review at a later date ... much later!

If the PHO is self-administered, it must have an experienced management team capable of truly communicating with both the physicians and the hospital administrators, and have appropriate data integration systems and report-generation abilities. This mix is difficult, as PHOs are newly defined organizations and the need for having an administrator with this unique combination of talents and pioneering vision is not easily found. Most hospital administration programs in accredited universities "churn" hospital administrators-to-be, with little clinical experience necessary to be able to empathize with professional staff. Most MCO administrators tend to be graduates of finance programs with little, if any, of the clinical background needed to assess quality and utilization numbers with any sort of epidemiological significance. This failure results in the inability to understand market opportunity and causes one to make flawed generalist assumptions. That, in succession, creates flawed contracting strategies and assumptions about market leverage.

Most physicians will tell you that if they wanted to be managers and administrators, they would not have gone to medical school.

When it comes to subcontractors, the PHO has the ultimate responsibility; therefore, it is critical to establish a budget and manage subcontractors such as psychotherapy, lab, imaging services, and physical medicine in strictest conformity with that budget. The same thing applies to subcontractors that work for you on the administrative side because of HIPAA regulations, privacy and security compliance, antitrust and contracting, anti-kickback and self-referral, medical records privilege, etc.

For example, Mary is married to Marvin the doctor. Mary works at your MSO or your PHO part-time every afternoon from noon until 5 p.m. as an independent contractor to act as referral coordinator. She quietly steers referrals to Marvin when the opportunity arises. What compliance and other issues might arise in this arrangement?

- Stark
- Anti-kickback
- Independent contractor rules
- State anti-self referral acts
- Unfair dealings/unfair advantage
- Workers' compensation regulations

Okay, so now you caught her doing it, by direct observation, by report analysis of referral patterns, however you do it. You decide to terminate her services "for cause."

Ummm, where is the set of standards against which "cause" for termination was based? Was it an attachment to her independent contractor agreement? Did she agree to abide by those standards of conduct?

So next, she files unemployment claims with the state. Wait; she was an independent contractor! Was she? Really? In the example, she worked from noon until 5 p.m. each day. Did she use your computers, your phones, your offices, or your software? Did she have to work according to the standards and rules of the organization? Hmmm … I wonder how the labor board will decide. Take out your checkbook, because if she is over 40, she will probably also file for wrongful termination, emotional distress, etc., etc., etc.

There's so much more to staffing a PHO than doctors seem to realize.

MANAGED CARE CONTRACTING WITH PAYORS

Often, MCOs will subcontract through the use of a network capitation model, whereby the subcontractor receives a prospective budget per-member-per-month (PMPM) and all claims received are paid against that pool of funds. If they contract directly with employers (also payors), the metric tends to be expressed in per-employee-per-month (PEPM). Based on current demographics in United States, that translates to about 1.6 lives per employee. Should the funds be more than enough to cover claims incurred, an incentive is received of more monies than the provider billed for each service. Should the services provided overrun the budget, a risk is deducted from the billed amount to offset the deficit. At no time is the established budget tampered with. This instills strong incentive for subcontractors to monitor their own utilization and outcomes for optimum performance under their capitated contract.

Exceptions as to what will be excluded for capitated responsibility are referred to as "carve-outs." When negotiating with IPAs*, if the PHO is small in size, it may be necessary to negotiate a carve-out from the PHO capitation and let the MCO deal with those subcontracted services. This writer feels that if the IPA will be full-service, it is wise to obtain the actuarial data and projections, as well as actual historical provider utilization reports and establish a budget for direct subcontracting.

GOVERNANCE ISSUES: CONTROL

Control is probably one of the most hotly debated issues in PHO management. Ideally, the PHO will have equal representation between hospitals and physicians, although the methods for voting and allocating ownership between physicians can vary greatly. Control is a major issue in life, why should it be different in business? All the participants should be satisfied with the representation and the decision process clearly defined and understood to allow for consensus that is binding and as timely as is needed. The law has some regulations about ownership and control.

Governance issues in a PHO are no party because of the myriad of PHO structures that we see throughout the country.

---

* Independent Practice Associations.

What do you call a PHO that purchases primary care practices? A PHO. What do you call a PHO that purchases practices and contracts with the former owner/physician as an independent contractor? A PHO. What do you call a PHO that only purchases the practices of retiring physicians and hires new ones as employees? A PHO. What do you call a PHO that purchases the practices and hires the former owner/physicians as salaried employees? A PHO. As I stated in Chapter 2 on IPAs, if you have seen one PHO, you have seen all PHOs.

So many of the other issues that arise for PHOs are the same for IPAs, and I will address them together in later chapters of this book. Such issues as credentialing, membership, and contracting will be covered together as there is little essential difference in the initial development stages.

MEDICARE ANTI-KICKBACK AND PRACTICE ACQUISITIONS

One area that I do want to address specific to PHOs is the Medicare anti-kickback issue related to hospital incentives to physicians. This problem has the potential to arise when a physician becomes employed by a hospital, the hospital purchases the practice, or gives special terms and conditions on loans, services, or office space among other things that could be construed as payment for the inducement of referrals, whether in cash or in-kind.

The Office of the Inspector General (OIG) has long been aware of a variety of hospital incentive programs used to compensate physicians either directly or indirectly for referring patients to the hospital. These arrangements are implicated by the Medicare and Medicaid anti-kickback statute*. Among other things, the statute penalizes anyone who knowingly or willfully solicits, receives, offers, or pays remuneration in cash or in-kind to induce or in return for

1. Referring an individual to a person for the furnishing or arranging for the furnishing of any item or service payable under the Medicare or Medicaid program, or
2. Purchasing, leasing, ordering, arranging for or recommending purchasing, leasing, or ordering any good facility, service, or item payable under the Medicare or Medicaid program.

Violators are subject to criminal penalties, or exclusion from participation in the Medicare or Medicaid programs, or both. In 1987, Section 14 of the Medicare and Medicaid Patient and Program Protection Act (P.L. 100-93) directed the Office of the Inspector General of the Department of Health and Human Services to promulgate (develop) "safe harbor" regulations. These "safe harbors" were published on July 29, 1991†.

In these relationships between physician and hospital, it is necessary to beware of certain activities that may prove suspect of violation, including the following:

---

* 42 U.S.C. Section 1320a-7b(b).

† 42 C.F.R. Section 1001.952, 56 Fed. Reg. 35. '452.

1. Payment of any sort of incentive by the hospital each time a physician refers a patient to the hospital;
2. Use of free or significantly discounted office space or equipment (in facilities usually located close to the hospital);
3. Provision of free or significantly discounted billing, nursing, or other staff services;
4. Free training for a physician's office staff in areas such as management techniques, CPT coding, and laboratory techniques;
5. Guarantees that provide if the physician's income fails to reach a predetermined level, the hospital will supplement the remainder up to a certain amount. (This one has some side rules that come into effect in healthcare manpower shortage areas. Check with an attorney well versed in health law.);
6. Low-interest, interest-free loans, or loans that may be "forgiven" if a physician refers patients (or some number of patients) to the hospital;
7. Payment of the cost of a physician's travel and expenses for conferences;
8. Payment for a physician's continuing education courses;
9. Coverage on the hospital's group health insurance plans at an inappropriately low cost to the physician; and
10. Payment for services (which may include consultations at the hospital) that require few, if any, substantive duties by the physician, or payment for services in excess of the fair market value of the services rendered.

One document floating around since December 22, 1992, a letter to T. J. Sullivan, who at the time was Technical Assistant at the Office of the Associate Chief Counsel of the Employee Benefits and Exempt Organizations of the Internal Revenue Service, from D. McCarty Thornton, then Associate General Counsel at the Inspector General Division, answers Sullivan's question regarding the IRS's views concerning the application of the above Medicare and Medicaid anti-kickback statute to certain types of situations involving the acquisition of physician practices by hospitals.

Attorney Thornton stated in his response:

> We have significant concerns under the anti-kickback statute about the type of physician practice acquisitions described in your inquiry to us. Frequently, hospitals seek to purchase physician practices as a means to retain existing referrals or to attract new referrals of patients to the hospital. Such purchases implicate the anti-kickback because the remuneration paid for the practice can constitute illegal remuneration to induce the referral of business reimbursed by the Medicare or Medicaid programs.

He goes on to further state that,

> "Since tax exempt hospitals are generally required to participate in the Medicare and Medicaid programs as a condition of obtaining or maintaining their tax exempt status, the anti-kickback statute is necessarily a significant issue to be addressed by them."

Further in the letter, he addresses suspicious or troublesome practices that might lead the officials to believe that there is an inducement by the way the deal is structured. He says,

> "The following are specific aspects of physician practice acquisition or subsequent activities that may implicate or result in violations of the anti-kickback statute. Our comments focus primarily on two broad issue categories: (1) the total amount paid for the physician's practice and the nature and type of items for which the physician receives payment; and (2) the amount and manner in which the physician is subsequently compensated for providing services to patients."

He further addresses the issue of physician compensation in his second footnote that reads,

> "We would also note that while the anti-kickback statute contains a statutory exemption for payments made to employees by any employer, the exception does not cover any and all such payments. Specifically, the statute exempts only payments to employees which are for 'the provision of covered items or services.' Accordingly, since referrals do not represent covered items or services, payments to employees which are for the purpose of compensating such employees for the referral of patients would likely not be covered by the employee exemption."

Thornton further elaborates on previous case law decisions in his letter to Sullivan where decisions have been made after scrutiny of the structure of the sale. He states,

> "Under the anti-kickback statute, either of the previous categories of payment could constitute illegal remuneration. This is because under the anti-kickback statute, the statute is violated if one purpose of the payment is to induce referral of future Medicare or Medicaid program business.'"

He addresses the necessity to scrutinize these deals, including the surrounding facts and circumstances to determine the purpose for which payment has been made. He states,

> "As part of this undertaking, it is necessary to consider the amounts paid for this practice or as compensation to determine whether they reasonably reflect fair market value of the practice or the services rendered, in order to determine whether such items in reality constitute remuneration for referrals. Moreover, to the extent that a payment exceeds fair market value of the practice or the value of the services rendered, it can be inferred that the excess amount paid over fair market value is intended as payment for the referral of program-related business."

To this point, he quotes a decision statement made in *United States v. Lipkin*, 770 F.2d 1447 (9th Cir. 1985).

To address the issue of the method of the valuation of the practice and fair market value, he says,

> "When considering the question of fair market value, we would note that the traditional or common methods of economic valuation do not comport with the prescriptions of the anti-kickback statute. Items ordinarily considered in determining the fair market value may be expressly barred by the anti-kickback statute's prohibition against payment for referrals. Merely because another buyer may be willing to pay a particular price is not sufficient to render the price paid to be fair market value. The fact that a buyer in a position to benefit from referrals is willing to pay a particular price may

only be a reflection of the value of the referral stream that is likely to result from the purchase. This deviation from the normal economic model was made expressly clear in the safe harbor provisions. For purposes of determining the value of space or equipment rentals, 'fair market value' is specifically defined to exclude the 'additional value one party … would attribute to the property (equipment) as a result of its proximity or convenience to sources of referrals or business otherwise generated.'"*

The reason that I reiterate so much of this letter is because I continue to see these deals happening throughout the country. The problem is so widespread that more than hundreds of special agents are routinely sent into the field from a joint task force formed in May 2009 by the U.S. Department of Justice (DOJ) and Health and Human Services (HHS) Health Care Fraud Prevention and Enforcement Action Team (HEAT), to uncover, first-hand, the actions, tactics, and strategies of these deals and other potentially fraudulent activities by physicians, hospitals, and others.

From talking to and working with physicians involved in PHO activities, I am acutely aware that many either are not aware, or even worse, once I make them aware, their attitude is, "Not my hospital; they would never do anything illegal." Or I hear, "It will never trickle down here to this little town." Get real! Fraud is fraud, and abuse is abuse, whether you do it in urban America or rural America! Wait! There is more!

Also, one little unrealized fact: if you have a problem and a consultant finds it, there is no consultant–client privilege. Scoot that consultant's contract under your attorney's door if you feel you will need privilege protection and caution them not to send discoverable notes, reports, or e-mails or voicemails without going through the attorney.

Thornton points out that,

"When attempting to assess the fair market value (as the term is used in the anti-kickback analysis) attributable to a physician's practice, it may be necessary to exclude from consideration any amounts which reflect, facilitate, or otherwise relate to the continuing treatment of the former practice's patients. This would be because any such items only have value with respect to the ongoing flow of business to the practice. It is doubtful whether this value may be paid by a party who could expect to benefit from referrals from that ongoing practice. Such amounts could be considered as pay-merits for referrals. Thus, any amount paid in excess of the fair market value of the hard assets of a physician practice would be open to question. Similarly, in determining the fair market value of services rendered by employee or contract physicians, it may be necessary to exclude from consideration any amounts which reflect or are affected by the expectation or guarantee of a certain volume of business (by either the physician or the hospital). Specific items that we believe would raise a question as to whether payment was being made for the value of the referral stream would include, among other things: (1) payment for goodwill; (2) payment for the value of the ongoing business unit; (3) payments for covenants not to compete; (4) payment for exclusive dealing arrangements; (5) payment for patient lists; or (6) payment for patient records."

Thornton says that any payments for the above are questionable when there is a continuing relationship between seller and buyer, and when the buyer relies on

* 42 C.F.R. 1001.952(b) and (c), 56 Fed. Reg. 35971–35973, 35985.

referrals from the seller, and that these arrangements raise grave questions of compliance with the anti-kickback statute. I love his closing statement in the letter; he says,

> "We believe that many of these arrangements are merely sophisticated disguises to share the profits of business at a hospital with referring physicians, in order to induce the physicians to steer referrals to the hospital."

Gee, that was to the point!

In any event, the PHO business has, and will continue to be an interesting strategy for hospitals and physicians and other ancillary providers of healthcare services. Clearly, the time for change has arrived. Everybody in healthcare is convinced that the current marketplace is serious about doing something to change the system as it currently exists. What the changes will be is still somewhat of a mystery. All provider types are rushing to position themselves in a posture that will enable them to accommodate the reform as it is mandated.

In Table 3.1 and Table 3.2 are two outlines that I use often to weave my way through the issues of PHO development, whether in the form of a for-profit or a not-for-profit structure, whether a standard PHO or a medical foundation. You may find it helpful as a checklist.

The reason I have not elaborated these items is because many would require a skilled health law attorney to address them appropriately. For me to address these items in detail would be skating too close to unauthorized practice of law.

In summary, hospitals that build PHOs from a self-serving standpoint proliferated in the 1990s. How did that work out for them? <Smirk>

The necessity for alignment and integration is key to the viability of a PHO. Building just another "51/49 thing" that has no alignment of incentives, solves no business dilemma, and just "has one to have one" is doomed to failure. A hospital without an engaged medical staff is called "real estate." There is one down the street from my house. It has over 200 beds. The new folks who want to buy it want to scrape it off the land and build a new mixed-use residential/retail combination site where the hospital once stood. The architects, engineers, and developers came to our neighborhood association meeting where I am on the land-use committee. They explained in detail the brown field issues associated with land that was once a hospital. It is not pretty, and they will "buy it for a song" compared to land values that were not a hospital in the past. This particular hospital simply moved to a new neighborhood. Other hospitals have simply passed into history for lack of medical staff alignment and engagement.

To determine if a PHO will be appropriate for a specific case requires thorough analysis. The fact that everybody is forming something does not imply that it is okay to follow carbon copies of what someone else did in the next town, the next state, or the next country.

Start by taking the two charts supplied in Tables 3.1 and 3.2, and send them out to the potential members who you plan to invite. Ask them to rank-order the importance or provide commentary on their preferences and why. This will save time and money. Otherwise, your consultant will have to perform this step for you before he or she can get anything else done. Be sure to use the charts, checklists, and appendices the same way. Get consensus. That is the first step toward alignment and organizational development.

TABLE 3.1
PHO Organizational Development Punchlist

Purpose/Description:

- Legal entity of MDs and hospital
- Facilitate contracting
- Improve management cost and use
- Create new healthcare resource

Management:

- Part-time versus full-time
- Hiring authority
- Reporting relationships
- Staff size

Legal Structure:

- For-profit
- Not-for-profit
- Taxable for profit
- Tax exempt
- By contract

Policy Issues:

- Contracting parameters
- Credentialing decisions
- Provider payment decisions
- Lock-in of providers:
  - Required to participate in every contract
  - Opt-out choices available
  - Exclusivity arrangements
- Utilization management policies
- Quality assurance
- Quality improvement policies

Focus of Activity:

- Managed-care contracting
- Direct-employer contracting
- Product development

Financing:

- Initial capital development
- Outgoing expense funding
- Hospital capital contributions
- Physician capital contributions

Ownership:

- 100 percent hospital
- 50 percent hospital, 50 percent MDs
- Other

Legal Issues:

- Antitrust
- Fraud and abuse

Continued

TABLE 3.1 (*Continued*)
PHO Organizational Development Punchlist

- Tax-exempt issues
- General taxation
- Corporate practice of medicine prohibitions in each state
- State insurance laws
- ERISA
- Tort law
- Securities law
- Errors and omissions coverage
- Officers and directors coverage

Governance:

- Composition of the board
- Appointment or election
- Decision making
- Committees

Legal Documents Necessary:

- Bylaws, articles of incorporation
- Service agreements
- Shareholder agreements
- Purchaser agreements
- Management agreements
- Antitrust compliance program
- Credentialing application
- Credentialing process
- Securities disclosure document
- Financing documents
- Private Placement Memorandum
- HIPAA Business Associate Addendum
- Job descriptions
- Employment policies
- Healthcare informatics
- Electronic medical record integration
- Meaningful use
- Interoperability
- Data security
- Data storage

Marketing:

- Website development
- Website hosting
- Logo development
- Sales and marketing collateral
- Broker and producer relationships

TABLE 3.2
Medical Foundation Organizational Punchlist

All the preceding (i.e., Table 3.1) PLUS:

Purpose/Description:

- Legal entity focused on the management and delivery of quality healthcare services

Legal Structure:

- Tax-exempt corporation

Ownership:

- MD Owned
- Hospital Affiliated
- Hospital owned

Governance:

- Corporation
- Decision-making process

Management:

- Background required resources needed reporting relationship

Focus of Activity:

- Practice management
- Managed-care contracting
- Research
- Ancillary services
- Purchasing
- Facilities development accounting
- Billing
- Personnel employment

Policy Issues:

- Practice evaluation of benefit plans
- Authority to contract exclusivity
- Participation criteria

Financing:

- Hospital support
- Hospital/MD support
- MD support
- Loans

If you do decide to hire a consultant, in your vetting process and interviews, ask what they know about the social science of organizational development, team building, and motivating teams. Very little of this process is tied to documents generation. Documents simply memorialize the policies and procedures you have developed and adopted. It makes no sense to borrow someone else's template. You can purchase a document set template for a few thousand dollars and never once be concerned with whether you survive or not, but you would have great documents!

Hopefully this chapter has provided some insight into the more common issues involved in the formation of a PHO. Beyond this, you will need an expert to evaluate your specific situation.

4 Accountable Care Organizations (ACOs)

ACOs create incentives for healthcare providers to work together to treat an individual patient across care settings—including doctor offices, hospitals, and long-term care facilities.

Both the 2010 Patient Protection and Affordable Care Act and economic realities are compelling hospitals and physicians to rethink how they deliver care. The current system, which generally rewards physicians and hospitals for "doing more," is no longer sustainable. New models and payment methods are being sought that reward providers on the basis of quality and cost effectiveness rather than volume.

Principal in buzzwords among these models are Accountable Care Organizations (ACOs). The Medicare Shared Savings Program and participating private payers will reward ACOs that lower growth in healthcare costs while meeting performance standards on quality of care.

Participation in ACOs is voluntary, and not all health facilities or physicians may be willing or able to adopt this model. To be an ACO, however, one does not simply declare it. The term is one that is used for a specific purpose, and an entity is not an ACO unless the Centers for Medicare & Medicaid Services (CMS) declares that it is. Otherwise, it really should be selecting another name and acronym.

In a press release dated December 19, 2011*, the U.S. Department of Health and Human Services (HHS) Secretary Kathleen Sebelius announced that thirty-two leading healthcare organizations from across the country will participate in a new Pioneer Accountable Care Organizations initiative made possible by the Affordable Care Act. The Pioneer ACO initiative will encourage primary care doctors, specialists, hospitals, and other caregivers to provide better, more coordinated care for people with Medicare and could save up to $1.1 billion over five years. Under this initiative, operated by the CMS Innovation Center, Medicare plans to reward groups of healthcare providers that have formed ACOs based on how well they are able to both improve the health of their Medicare patients and lower their healthcare costs.

Secretary Sebelius stated that she was excited that so many innovative systems elected to participate in the initiative, and she explained that there are many other ways that healthcare providers can get involved and help improve care for patients. The Pioneer ACO initiative was just one of a menu of options for providers looking to better coordinate care for patients and use healthcare dollars more wisely. The Pioneer ACO model is designed specifically for groups of providers with experience

* http://www.innovations.cms.gov/documents/pdf/PioneerACO-Press_ReleaseFINAL_12_19_11v3_2.pdf.

working together to coordinate care for patients. The Medicare Shared Savings Program and the Advance Payment ACO Model, both announced in October 2011, are also ACO options for providers. In fact, as of October 2011, there were four models from which a group of providers could choose.

These models currently include, but may not be limited to the following.

1. *Medicare Shared Savings Program for Accountable Care Organizations:* The Medicare Shared Savings Program will allow providers who voluntarily agree to work together to coordinate care for patients and who meet certain quality standards to share in any savings they achieve for the Medicare program. ACOs that elect to become accountable for shared losses have the opportunity to share in greater savings. ACOs will coordinate and integrate Medicare services, with success being gauged by roughly thirty quality measures organized in four domains. These domains include patient experience, care coordination and patient safety, preventive health, and at-risk populations. The higher the quality of care providers deliver, the more shared savings their ACO may earn, provided they also lower growth in healthcare expenditures. The Shared Savings Final Rule[*] is located online at http://www.regulations.gov/#!documentDetail;D=CMS-2010-0259-1591 [doi, January 7, 2011].
2. *Advance Payment Accountable Care Organization Model:* The Advance Payment model will provide additional support to physician-owned and rural providers participating in the Medicare Shared Savings Program who also would benefit from additional start-up resources to build the necessary infrastructure, such as new staff or information technology systems. The advance payments would be recovered from shared savings achieved by the ACO.
3. *Pioneer Accountable Care Organization Model:* The Pioneer model is an initiative complementary to the Medicare Shared Savings Program designed for organizations with experience providing integrated care across settings. The Pioneer model tests a rapid transition to a population-based model of care, and engages other payers in moving toward outcomes-based contracts. The initial group of Pioneer sites, slated to be announced later this year, will be positioned to rapidly demonstrate what can be achieved when we provide highly coordinated care to Medicare fee-for-service beneficiaries.
4. *Financial Models to Support State Efforts to Integrate Care for Medicare-Medicaid Enrollees:* A longstanding barrier to coordinating care for Medicare and Medicaid enrollees has been the financial misalignment between Medicare and Medicaid. This initiative will test two models—a capitated model and a managed fee-for-service model—for states to better align the financing of the Medicare and Medicaid programs and integrate primary, acute, behavioral health, and long-term services and supports for

[*] Federal Register, Volume 76, Number 212 (Wednesday, November 2, 2011), Rules and Regulations, pages 67802–67990.

> Medicare and Medicaid enrollees. For those states that are interested in testing these two models, CMS is offering streamlined approaches and technical assistance to support necessary planning activities.

The problem here is that we now have at least four of perhaps many more models to be proposed or deployed, all described by the same high-level three initials. This is confusing and will mire the outcomes, terminology used by the media and the market, and cause providers, statisticians, and the public to become confused by the lack of clarity and ambiguity in the use of the terminology. Few refer to the organizational form by saying, "We are building a Pioneer Model ACO." They simply say, "We are building an ACO." The listener nods, parks that piece of data into a file in his brain (potentially the most inaccurate file), and goes on to develop the basis of opinions, observations, and perhaps reactions and responses that could then be totally flawed.

For the record, I am not a believer in this program as it is currently designed and deployed, and I have little faith in its ability to complete its mission. My objection and skepticism is not political in nature; it arises from deep experience in hands-on development and operation of several local and regional integrated health delivery systems and now as CEO and founder of the world's largest, single, independent, NGO, globally integrated health delivery system®, a term of art used to describe not only physician and hospital and administrative integration in the United States and ninety-five other countries, but also alignment of incentives for hospitals, providers, employers and unions (payers), investors, and patients.

I know from experience what it takes to develop successful integrated groups, and run them, and measure their success and compare outcomes across disparate entities. Without paraphrasing the entire ACO regulation playbook (as currently published) and including it, with my point-by-point objections and observations, I simply do not see the necessary guidance or infrastructure in place from the CMS or HHS, despite all its current publications* or a recipe for how to go about building and operating a successful entity that will stand up for three years, the period for which the contract is awarded from CMS.

I also know firsthand the myriad ways in which political and pedantic physicians (PPP) in many entities will get in their own way and derail the entity that can identify patients and say, "I did this because my patient is "S-O-U-Pier" (Sicker, Older, Uglier and Poorer, as the old saying goes) and so this is why I did it my way" (cue the Sinatra music). In the ACO, neither the doctors nor the patients themselves are known to the physicians; there is actually the prevention of a "Hawthorne effect."†

In fact, Medicare beneficiaries will not receive advance notice of their ACO assignment. However, providers participating in ACOs will be required to post signs in their facilities indicating their participation in the program and to make available

---

* The 429-page proposal was placed on public display on March 31. It provides a 60-day comment period and was released for publication in the April 7, 2011, Federal Register. It was then replaced by yet another document on October 20, 2011, when it released its Final Rule.

† A term referring to the tendency of some people to work harder and perform better when they are participants in an experiment. Individuals may change their behavior due to the attention they are receiving from researchers rather than because of any manipulation of independent variables.

standardized written information to Medicare fee-for-service beneficiaries whom they serve. Additionally, all Medicare patients treated by participating providers must receive a standardized written notice of the provider's participation in the program and a data use opt-out form. That is tantamount to preparing a banquet without RSVPs and simply preparing perishable foods that perhaps too few come to eat. I say too few because in order to have statistics, one must have a significant enough sampling of data. It is also sort of creepy. As a Medicare beneficiary, I would probably ask to opt out.

Oh, but they addressed that too: An ACO must have at least 5,000 beneficiaries. If an ACO accepted into the program falls short of the 5,000 requirement, it will be placed on a corrective action plan. How does one correct an action if one knows not who is in the sample group? And please do not quote the propaganda from the publications to me; tell me "how" it will be done in reality, from a practicable standpoint.

Under the Final Rule, a group of providers and suppliers of services agree to work together with the goal that patients get the right care at the right time in the right setting. The Final Rule requires that each group of providers be held accountable for at least 5,000 beneficiaries annually for a period of three years. Has the government ever heard of the snowbird effect? Growing up in South Florida, every November, the traffic jams would start along I-65, I-75, and I-95 as seniors and Canadians would flock like geese to South Florida and remain there until about April. They also headed to Texas, Arizona, Las Vegas, and California. Now that I have knowledge about expatriate markets, I also know they flock to expat colonies in Mexico, Portugal, Spain, Italy, and Central American countries such as Costa Rica, Ecuador, and Panama.

So, how will doctors be held accountable for these folks' care annually for thirty-six months?

Here is another weird and seemingly manipulative way of looking at the data: Medicare fee-for-service beneficiaries will be retroactively assigned to ACOs based on primary care utilization during a performance year. "CMS are proposing to assign beneficiaries for purposes of the Shared Savings Program to an ACO if they receive a plurality of their primary care services from primary care physicians within that ACO."

They must have altruistic geniuses as government statisticians with some of the most sophisticated statistical modeling formulae in the world; and working for government wages and benefits. So, I ask: If they retroactively can assign Medicare beneficiaries, why should ACOs be bothered with the time, expense, and complexity to develop and deploy a corrective action plan? Where does the money come from to fund development, deployment and operation, and measurement of effectiveness or feedback of said corrective action plan?

The Final Rule also links the amount of shared savings an ACO may receive, and in certain instances shared losses it may be accountable for, to its performance of (1) quality standards on patient experience (largely subjective); (2) care coordination and patient safety (that they cannot control); (3) preventive health (that they cannot control, as we have tort laws that address battery if we treat patients without consent—so don't go jabbing that flu shot in their arm without consent to make your ACO numbers!); and (4) at-risk populations.

In the draft ACO regulations, CMS established a single ACO risk adjustment factor using historical data from all participants, and did not modify it throughout the participation period. This policy reflected CMS' concern that risk scores would increase due to changes in coding (which sounds as if CMS suspected there would be naughty manipulative doctors and administrators with time to do this for some anticipated gain) rather than actual changes in patient status, which would inappropriately lower the benchmarks and allow additional "savings" to be realized. Therefore, the risk score of the initial group of members would be applied throughout the participation period, regardless of changes in the actual membership or changes to the health status of those members.

Potentially, this approach would discourage ACOs from recruiting the "S-O-U-Pier," higher-severity patients who would be the greatest beneficiaries of an integrated and coordinated care approach. The Medicare Payment Advisory Commission (MEDPAC) even suggested that this approach "would create incentives for ACO providers to encourage existing patients who are costly to seek care elsewhere"—a polite way of describing "patient dumping" we so often experienced in the early 1990s' days of capitated reimbursement from HMOs. Another challenge could arise from changes in a patient's condition, for example, discovering a previously undiagnosed cancer, which would not be accommodated.

In the Final Rule, CMS significantly revised its approach to risk adjustment, noting that "commenters have persuaded us of the need to better account for risk associated with changes in the ACO's beneficiary population," although it remains concerned about the effects of upcoding the diagnoses used in the HCC-based* risk scoring methodology. While it did not adopt the approach used by Medicare Advantage plans—namely the Risk Adjustment Processing System (RAPS)—it has taken steps to deal with the issues raised by the commenters.

For newly assigned beneficiaries, CMS will now annually update the ACO's risk score to adjust for the severity and case mix. This will accommodate higher-severity members who may join the ACO (by utilizing its PCPs [Primary Care Physicians]) to benefit from the care coordination services that the ACO should offer. This will give the ACO the incentive to recruit such patients. But, how will they recruit people to be in the ACO if neither the beneficiary nor the physician knows who is in the ACO measurement population?

For continuously enrolled members, the methodology is somewhat less clear. CMS will compute HCC risk scores for these members but will only adjust the member's health status if the scores decline. The rule is silent on what happens if the scores increase, so presumably there is no adjustment. The only adjustment will be for changes in age: as the population grows older, its risk score will increase. This accommodates several of the problems, while leaving others still in place. Again, I argue that the infrastructure is not there to make this successful over the three-year period.

* Hierarchical Condition Categories. The Medicare risk adjustment payment system uses clinical coding information (HCCs) to calculate risk premiums for Medicare Managed Care Organizations (MCOs). On average, 80 percent of the codes in the Risk Adjustment Processing System (RAPS) come from claims submitted by a physician. The claim is really a proxy for a medical record entry, and the assumption is that there will be documentation in a form acceptable to CMS for every code submitted in RAPS.

As a result, for this author, right or wrong, I hereby state with the courage of my convictions that I have little to no faith in the success of ACOs between 2012 and 2016, and hereby close this introduction and opinion section. If I am incorrect, so be it. As a reader and purchaser of this book, if I am indeed wrong, I will accept your criticism with respect, grace, and dignity.

For the record, what I do respect and hold in high regard are the philosophies surrounding the ACO concept, including

1. Doing the right thing by patients because it is right to do so
2. An approach that includes continuity of care when feasible and practicable
3. Cost containment through aggressive prevention and cost avoidance
4. Disease management for at-risk populations (health access disparities, limited English proficiency, Lesbian, Gay, Bisexual, Transgender (LGBT) patients, those that are emotionally disturbed, mentally ill, suffer from dementias, neurologically impaired, mentally retarded, physically challenged by paralysis conditions, and musculoskeletal disorders that confine them to wheelchairs or are housebound and have mobility challenges, etc.)
5. Critical assessment of clinical and patient satisfaction outcomes
6. Clinical and economic integration and alignment of payors, providers, and patients

What I feel is missing from the ACO programs is a structured approach to designing an analytic infrastructure to collect, normalize and consolidate, and analyze data across patient populations, demographics, utilization, risk scores, costs, and other factors that are beyond the control of the organizers or participants of the ACO, because their patients maintain free will, freedom of choice, freedom of geographic diaspora*, and the freedom to care or not care about their own health regardless of what the doctor orders. All of these critical elements will not be easily managed but will have a direct effect on the financial and clinical success and reputation of the ACO.

These analytics are critical in importance in the formative stages of any integrated health delivery system because the strategic framework must be determined first, and the outcomes report design must be decided upon prior to the development of the database that will be used to collect and maintain the data and produce the reports. As physicians will be recruited into the group to make the 5,000 patient minimums, little if any consideration may be paid to their patient mix, risk and outcomes scores (who measured them, what was measured, and how were the calculations done, for starters), and their coding habits and costs to payors. Without this framework or set of known characteristics (an integral part of the vetting process that is not contained in many credentialing and privileging applications), the vision of the future for each ACO will be more difficult to portray, share, and use as a road map to the future.

The lack of analytics in the 1990s contributed to the demise of many IPAs (Independent Practice Associations), PHOs (Physician Hospital Organizations), MSOs (Management Services Organizations), and Clinics Without Walls, and many older, wiser physicians, who played the last round, will be reticent to repeat this

* A dispersion of a people, language, or culture that was formerly concentrated in one place.

experiment at the latter stages of their careers when their last shot at gathering wealth for support during retirement is at the forefront of their decision to participate. They define "insanity" as doing the same experiments over and over, without the ability to obtain and analyze meaningful data or change circumstances.

In order to be able to establish a framework and a strategy for an ACO, the entity will have to designate someone who can understand the ACO contract, including major factors such as medical costs, risk scoring methods, out-of-ACO utilization, and other challenges consistent with managed care and government contract interpretation. Furthermore,

- Who will do this?
- What do they need to know?
- Where will they learn it?
- How will it be taught?
- By whom?
- What are the credentials, training, and experience to qualify the instructor?
- Who will pay for the education and costs to learn this?
- What if mistakes are made; who will suffer the consequences?

Most managed care contracts do not have an initial term of three years. In fact, most of my esteemed colleagues would advise against doing so with an unknown payer with an unknown track record. Yet, CMS/HHS is a familiar payor offering this contract. This payor has a track record of making amendments in the form of revised final rules, written by policy makers and politicians, instead of people who have to carry out the changes, and just sending out the Federal Register each week to let you find "serious stuff you need to know."

Next, the contract with the government between the ACO and CMS/HHS is not the only contract of concern. At a deeper level, there will be some form of provider agreement between primary care physicians (PCPs) and specialist physicians (SPs).

This is a "cart before the horse" challenge. In previous CMS demonstration projects, CMS has provided the ACO with the underlying claims data for the participants in the demonstration. In the ACO, CMS must initially develop the methodology to assign patients to their respective PCPs. Before that assignment can be carried out, the ACO must identify the PCPs who will participate in the ACO. This entire process can take months, because, for starters, in accordance with National Committee for Quality Assurance (NCQA) credentialing and recredentialing guidelines, an entity can take up to eighteen months to complete credentialing on a physician. ACOs that are formed as a separate corporation are required to perform credentialing via primary source verification, or delegate the task to some other entity that can do it for them under contract. A cost is involved with this task, and it is not inexpensive. ACOs cannot wait eighteen months for this. They will require some ability to analyze data initially from that data which is currently available, which might provide some guidance on future metrics required to evaluate and manage the activities of the ACO.

Another little glitch comes from the Health Insurance Portability and Accountability Act of 1996 (HIPAA), Public Law 104-191. If the PCP has no current contractual relationship with the ACO, then HIPAA exclusions such as Treatment, Payment and Operations (TPO) waivers do not apply. Therefore, the PCP will be prevented from releasing any data to the ACO. The ACO and the PCP could use a "clean agency" that enters into a business associated agreement with the PCP, who will then perform the analysis and provide the ACO with a summary report that does not release any protected health information. But, will that data be representative of those who will be assigned through the ACO by CMS? Why would one assume so? This is business, contracting, liability for breach, financial viability, and not Occam's Razor*. Reliance upon flawed assumptions could prove disastrous.

Next, hospitals have their own data, which is a combination of inpatient and outpatient services, as well as services provided by ancillary facilities that the hospital may own or bill for, such as a hospital-owned sub-acute facility or home health agency. Hospitals may also possess data for billings by physicians who are employed by the hospital, although except in rare instances in the United States, most hospitals will not be the employer of record of all the ACO physicians. Therefore, the usefulness of the data is limited and essentially locked up in that which I refer to as "sacred silos."

Merely the extrapolation of the data from the sacred silos is a challenge because few ACO administrators have the expertise to dig through disparate claims payment systems and sort the data. Again, where will they obtain the time to do this? The training? Who will pay to have it done as an initial capital investment into the development of the ACO? To me, it is a great exercise for a consulting firm to identify a client base but for which, in reality, there isn't a client out there to pay for this to be done. A product that answers a need and solves a problem in the market, but has no customers often fails for the following three reasons: No budget, no time, and no identification of the need yet but those who don't know what they don't know.

Additional expertise of third-party payment systems will be necessary for this analyst. As the leading trainer for managed care contracting and one who presents more training opportunities than any other association, institute, or conference company on managed care training, and essentially the only trainer that incorporates hands-on training for these techniques, worldwide … neither the thirty-one ACOs who have been contracted by HHS/CMS nor those who are contemplating formation have sent a candidate to learn this to one of my courses.

The analyst will need to understand third-party reimbursement for non-hospital providers and have specific knowledge of claims payment systems. Additional clinical nurse analysts and physicians will be needed to help evaluate details of the services provided and to answer questions related to appropriateness of pharmaceuticals, treatment approaches, medical necessity, and effects and implications of co-morbid conditions, together with actuarial expertise with population-based risk assessments and clinical management to refute potentially damning findings from the CMS' risk adjusters and altruistic statisticians I mentioned earlier. They will also have to have a well-developed acumen with IBNR reserves and other insurance-risk factors. I doubt that there are thirty-one "someones" looking for a job who have

* A principle that generally recommends that, from among competing hypotheses, selecting the one that makes the fewest new assumptions usually provides the correct one, and that the simplest explanation will be the most plausible until evidence is presented to prove it false.

that skills combination; in fact, I would doubt that there are thirty-one candidates out there who possess these skills in a single person who could delegate and manage others who might carry out the analytics, pull the results together, and form an assessment and create an action plan, and then "manage up" to the board of the ACO and its PPP physicians. To me, for an effective analyst/manager, that job would be worth, in today's (January 2012) market, at least $200,000 per year, plus employee costs and benefits. Which ACO is ready for this, and how will they recruit and screen applicants for the role? An ineffective or incompetent analyst could realistically cost the organization five to ten times that salary and employee costs in one year if the job is done wrong or poorly. Now that is what I call RISK!

If the ACO cannot manage the data and the analysis, it should not bother to collect, attempt to normalize, or report on that which cannot be measured, monitored, or continuously administrated over the course of the contract.

Finally, some ACOs may be associated with larger health systems that might participate in risk-based contracts or have existing IPAs, PHOs, or MSOs and operate some sort of centralized billing or claims management systems. These have been around for years, since the old days of Fred Rothenberg and EZ-Cap, which was acquired by QuadraMed and now is represented by MCI Healthcare. I am sure there are now other competitors out there and I have not evaluated very many of them in recent years.

First used in the early 1980s during the infancy of managed care, EZ-CAP was a first mover in the software programs that helped IPAs, PHOs, and MSOs manage and administer capitated contracts. Over the years, it has been installed at hundreds of clients worldwide and boasts the industry's most loyal customer base. According to marketing materials from MZI Healthcare, more than 35 percent of their clients have used the system for more than ten years, and new users allegedly continue to migrate to EZ-CAP from other vendors.

If that many health systems utilize the software, surely there is some repository within many health systems that could be accessed to obtain some of the data needed to build initial high-level strategies. The reality challenges in this case will include, among other practicable things, hospital administrators' perceived exposures and threats to physician alignment efforts by the nonemployed physicians who maintain privileges at multiple hospitals in a market, each of whom is being courted by each and every local competing health system to participate in their ACO exclusively.

Also, simply because data exists does not give one the right to assume that it is accessible, correct, or will be useful in this reformed reimbursement and operational environment.

5 Management Services Organizations (MSOs)

A Management Services Organization (MSO) is an organization owned by a group of physicians, a physician–hospital joint venture, or investors in conjunction with physicians. In some cases, the hospital owns the service bureau that sells various management services to medical staff. In other cases, it can act as a quasi third-party administrator and adjudicate and reprice claims and issue payment on behalf of self-funded employers, unions, and even insurance companies.

MSOs generally provide practice management and administrative support services to individual physicians, Independent Practice Associations (IPAs), Physician Hospital Organizations (PHOs), or small group practices. One function of MSOs is to relieve physicians of nonmedical business functions so that they can concentrate on clinical activities and attention to patients.

MSOs can act as Group Purchasing Organizations (GPOs) to help members of IPAs, PHOs, and group practices purchase their services more economically instead of individually, to achieve economies of scale. These cost savings may be passed on to physicians, who may use this cost advantage when negotiating with health plans and healthcare purchasers. In other cases, it could be one way that the MSO sustains itself financially by keeping those savings to run the organization as a co-op.

In some cases, MSOs simply provide a variety of business services to providers for a fee according to a fee schedule. In other cases, MSOs purchase the tangible assets, such as buildings, equipment, and supplies, and hire the staff as a Professional Employer Organization (PEO), of their client physicians and lease these assets back to the physicians. In these situations, the physicians continue to own their own medical records and health plan contracts and continue to practice in their own offices. This arrangement relieves physicians of yet another nonmedical aspect of running a practice.

In several cases, MSO have been able to develop programs to achieve malpractice discounts, discounted equipment leasing, shared staffing and benefits as a Multiple Employer Trust (MET), electronic billing, and electronic medical records (EMR).

One advantage that MSOs deliver is the ability to develop clinical guidelines and care standards for the participating member practices, thereby meeting clinical integration definitions and also being able to harvest a shared savings relationship or pay-for-performance program with third-party payors.

THREE BASIC ELEMENTS OF MSOs

Management Services Organizations (MSOs) should have three basic elements as part of their relationship with providers and ancillary providers:

1. The MSO provides management services for the providers and other users, as well as access to capital for expansion. This relieves the provider from being distracted from patient care by assuming the day-to-day centralized management functions of the practice, including capitation, contract management, marketing, utilization management, continuous quality improvement activities, group purchasing, and support services such as transcription, human resources development, billing, collections, and claims management services.
2. Affiliation with an MSO allows reorganization among the group to provide a vehicle for economic integration without blending the practices into one unified corporation, thereby allowing an IPA, or several affiliated IPAs, to affiliate with one MSO. Governance, real estate, compensation, and quality improvement issues can be addressed in an organized fashion, preparing the group for its future relationships with capitated managed care and facilitating direct contracting with employer-sponsored healthcare coalitions.
3. If desired by the providers, the MSO may purchase certain assets of the medical practices, providing equity in the practice that might otherwise not be liquid to the provider. This activity provides a one-time benefit to the providers in a group that might enable them to realize a return on the value of the ongoing business unit that they developed if they left the practice on other terms. This third element is not a requirement, however, of MSO development.

Often, we see providers who might not otherwise look for a buyer in times of financial uncertainty such as these, with dwindling profit margins and escalating overhead to gross ratios, look to anyone who will buy the practice as a "bailout" mechanism. These are usually solo practitioners who cannot see a brighter future ahead unless they forge ahead with the "fire sale." They are often exhausted, frightened, frustrated, have lost the opportunity to see beyond the immediate situation, have waited too long to "partner up" with other solo practitioners, and see no other way out but to sell the practice to the first entity that will buy it.

Faced with the decision to move ahead with MSO affiliation, the providers need to move to the next step of a "make-or-buy" decision. MSOs come in a variety of arrangements. The best arrangement is for the providers to make their own MSO or choose vendors who can outsource pieces of the MSO functions if they can handle the capitalization. This is usually in excess of $1 million for start-up and approximately $2 to 4 million for first-year cash requirements.

In this arrangement, the providers have full ownership and responsibility for the success by the management team that answers directly to them. The provider-governed Board of Directors makes all final decisions on contracting issues, allocation of revenue, and risk and asset management, while the staff carries out the decisions of the board as provider advocates with no other agendas. Only a fully owned model can guarantee this advocacy. Other options include rental of an existing MSO operated by other parties, such as management teams, who supply contractual services but offer no equity, or sharing in the development of an MSO with a hospital that may require management control and decreased autonomy in ambulatory care services management.

Often, hospital administrators who may have never administrated a provider practice feel that their management expertise and experience in hospital systems management transfers over to management of the day-to-day operations of a small, nondepartmentalized business unit. Typical matrix management and top-down management styles and the teachings of the "three-legged stool" known so well by all health administration/hospital administration M.H.A./M.B.A./M.P.H. candidates do not apply or transfer to medical group administration. Understanding of provider patient-centered concerns, market differences, customs, and values must be taken into consideration when evaluating management style and philosophy. Medical groups cannot be managed successfully as a hospital "department."

Another alternative might be to joint-venture with venture capitalists willing to co-capitalize an MSO with the providers. Experience shows that these individuals have strong ties to bottom-line performance and may not be willing to merely be silent partners with a checkbook. The providers' values, concerns, and customs may be sacrificed in this arrangement, thereby thwarting the cultural development and socialization that is so vital to the success of the organization. One possibility might be to co-venture with the chosen ancillary providers who may participate in the integrated healthcare delivery continuum, as these small businesses also will be required to participate in the centralized billing, capitation, and sub-capitation carve-outs but may not be able to fund their own MSO venture. Careful planning and development is required in this model so as not to create any regulatory violations concerning Medicare–Medicaid anti-kickback issues and Stark II concerns. The group must seek an experienced health law specialist for this task.

Undertaking due diligence in the evaluation of partnering the development of an MSO is critical to the success of the entity. Time must be taken to make a carefully thought-out decision, as divorce in this situation does not come without great emotional and financial expense.

The medical group must be able to evaluate its current governance structure in order to successfully implement the reorganization of its affairs to allow for coordination of the operations management in the MSO relationship.

Provider-independent contractor agreements may need to be revised to allow the MSO to serve as a messenger-model entity for the purpose of contracting with managed-care payors. The governance structure and bylaws of the organization must reflect the intent of how the organization will conduct its business affairs. Careful consideration must be given to the selection of leadership, how providers are credentialed into the group, and "right-sizing" for contracting efficiency. The MSO stands as its own entity and, as such, must respond in a representative capacity to the medical group. Day-to-day business operations of the medical group performed by the MSO must be addressed in a forum that will contain selected representatives of the medical group in a manner that preserves medical group advocacy.

The role of stock ownership in the medical group should be reexamined and modified to mean owning stock and voting rights to elect a Board of Directors and Officers, not significant ownership in assets. A Shareholder's Agreement should be prepared to reflect decisions made in this area. Experience has taught us that members should vote their membership democratically and not their shareholdership in a one-person/one-vote system to balance power.

To implement an MSO, the medical group must first begin planning and preparing on the administrative and provider relationship levels. The administrative staff should identify all the contractual and other business relationships that will be assumed by the MSO once the entity is up and running. The medical group will have to contact all lenders, vendors, and other interested third parties, as may be necessary, early in the process. Obtaining consents from third parties, especially in the area of managed-care contracts, is a time-consuming process. Other activities that will be centralized, such as group purchasing, waste disposal, transcription, equipment leases, insurance policies, etc., must all be reviewed. Some third parties will need sufficient notice to review files and process a payoff or assumption request and/or negotiation.

One difficult task is to effectively organize the individual providers to accept the management services concept. The old phrase "If you have seen one MSO, you have seen one MSO" is very true. When speaking of investment in an MSO, most providers will listen to plans about how to diversify revenue and maintain a strong position with managed-care payors. But telling them that they have to bill collectively and send all their billings to the MSO sends them into a panic. Regular communication efforts, both formal and informal, must be made to the medical group to foster involvement and support for the project. Meetings with provider staff members are an important aspect of the process. Individual concerns must he addressed openly and honestly, with the group's leadership being prepared to address all concerned providers on a one-on-one basis.

The leadership of the group should be prepared for the reality that some of the providers will not find the MSO affiliation an acceptable means of doing business. Although every effort must be made to maintain the core group, morale issues may develop if easy exit arrangements are not available for those providers desiring to leave the group. The rapidity of change may be more than some can handle. After all, to date there are few HMOs in Alaska, so they will have some place to carry out their lives and practices in a state of "ostrichosis."

The MSO relationship will require a management agreement that specifies how the MSO relates to the medical group, what services the MSO provides, and the governance mechanism by which a business partnership is created between the medical group and the MSO. Here is where the provider ownership of the MSO is most favorable. Will the MSO serve as contractor and manager of the providers with its own agenda, or will the MSO manage the business dealings of the providers as an advocate for the providers with a provider-driven agenda? The MSO should not interfere with the active practice of medicine. The medical group must look to the MSO expertise to provide business advice and coordination of the business aspects of the practice of medicine. Highly trained, exemplary administrative and clerical employees of the medical group become employees of the MSO, while technical employees will likely remain with the medical group because of federal and state regulatory compliance issues, such as OBRA '93 (Stark II), which requires employees used in ancillary services to remain under the direct supervision of the medical group rather than the MSO. Future anticipated developments in furthering the intentions of OBRA '93 will most likely discourage consolidation of ancillary services at the MSO level. With strong combinations of management expertise and patient-centered quality medical care, the group is destined for success.

Much of what I previously state for IPAs and PHOs also applies to MSOs. In fact, one could strip out everything that I mentioned that was of a nonclinical nature in the chapters of both IPAs and PHOs and even the ACO chapter, and if it has an administrative component, that part can be moved over to the menu and purview of the MSO. Therefore, for the sake of brevity, I elect not to repeat these points in this chapter. What I intended to do in this chapter was convey the "what." The "how" is spread throughout the preceding and following chapters and appendices. The "why" will become self-evident.

Section II

Integrated Health Delivery System Development

6 Corporate Form
Myriad Choices

GENERAL PARTNERSHIPS

A general partnership is a voluntary association of two or more individuals for business entities who agree to work together for a common business purpose. The partners, who own the business, share their profits or losses equally or as otherwise provided by agreement. Like proprietorships, partnerships can be formed easily. No formal steps are required to establish or maintain a partnership. In order to confirm certain aspects of their relationship, many partners enter into a written partnership agreement to specify their understanding (a meeting of the minds) on matters such as how the profits and losses of the business should be allocated, procedures for admission of new partners (stock purchase/entry documents), and withdrawal of existing ones (exit arrangements documents).

Partnerships offer business owners "pass-through" tax treatment. The partnership does not pay income tax. Instead, the income, profits, losses, and expenses of the partnership flow directly through to the partners, who then report their allocable share of income and expenses on their personal tax returns. Because partnership income is not taxed at the partnership level, operating a business as a partnership is, like a proprietorship, attractive from a tax perspective.

DISADVANTAGES OF GENERAL PARTNERSHIPS

Personal liability is the biggest problem with partnerships. In a general partnership, each partner is personally liable for all debts and obligations of the business. The consequences of this rule in a worst-case scenario could prove disastrous. If one partner innocently (or negligently) makes a mistake, it could cost all the partners. The problem is especially significant because individual partners may be forced to satisfy the business obligation out of their personal assets if the business assets are inadequate.

What about when someone leaves the partnership? A big problem with partnerships is that they lack business continuity. Without an agreement to the contrary, whenever an existing partner ceases to be a partner, whether as a result of retirement, death, expulsion, or the like, the partnership ordinarily is deemed to have been dissolved as a matter of law. Steps could be taken to continue the business, with or without one of the successors or heirs, if any, but the process is not automatic. If the heirs or successors decide not to continue the business, they could require a distribution of partnership assets and force a liquidation of the business. Talk about job security for attorneys and accountants! Given the volatile emotions and personalities in our healthcare delivery force, this is probably not the best way to go for a large group of corporately unrelated and related physicians in the form of an IPA or PHO, and it certainly would spell expensive doom for a true MSO of independent participants.

Lack of investment flexibility. General partnerships are financed either through capital contributions made by partners or by the use of debt. Most often in physician groups, the capital call is through personal funds. Rarely have I seen physician groups begin with a loan signed by the partners because they just do not start out a group endeavor like this with outside debt. Hence, many of these initiatives are sadly undercapitalized. I believe more than anything else that unless they have an MBA, they have not had the training and experience in business endeavors to know their options, nor do they each have the time or the inclination to approach their banker for money for a start-up that has no business plan, no corporate form (yet), no strategic market assessment (yet), or a good way to verbalize what the group's vision of the organization will be (yet). Although there is some flexibility in financing a general partnership, it is often less than that available in other business forms because a decision to bring in a new partner to raise needed money usually requires allocating some management responsibility to that partner.

LIMITED PARTNERSHIPS

Advantages of Limited Partnerships

Another form of business is the limited partnership, which is a distinct legal entity created under state law. Every limited partnership has at least one general partner who manages the limited partnership, and at least one limited partner. Limited partnerships can raise money from limited partners without conferring management responsibility as well. Therefore, limited partners have very little say in managing the corporation. In exchange for their limited control over partnership affairs, the limited partners have limited liability for partnership obligations. They are at risk only to the extent of their investment in the partnership. The general partner, in contrast, has unlimited liability for the partnership's debts and obligations. A limited partnership is formed by filing a certificate of limited partnership with the designated state agency, typically the Secretary of State. The partners can structure their relationships with each other and address certain business issues through a limited partnership agreement.

Pass-through tax treatment. Although the tax treatment of limited partnerships is extremely complicated, if appropriate steps are taken, the income and losses of the business will, like those of proprietorships and general partnerships, flow through to the individual partners in accordance with their partnership shares via a K-1 document filed with each investor's tax return. This structure avoids the double level of tax caused by a tax on the business itself.

Financial flexibility. Limited partnerships offer more financing opportunities than general partnerships because the former provide a vehicle for raising money (from limited partners) without having to take in the investors as general partners.

In some instances, financing techniques are simpler to implement than those techniques available to corporations.

Limited liability of limited partners. In a limited partnership, a limited partner, whether a person or a business entity, is personally at risk for the obligations of the business only up to the amount invested in the partnership. The party's liability has been limited, and personal assets cannot be used to satisfy the obligations of the business. In exchange for this limited liability protection, however, a limited partner is given a minimal voice in managing the partnership's affairs. A limited partner who participates in the management of the business in a more meaningful way will be stripped of this limited liability.

Disadvantages of Limited Partnerships

General partner liability for business debts. Unlike limited partners, the general partner in a limited partnership is fully liable for the obligations of the business. In order to insulate themselves from unlimited personal liability for the debts and obligations of the business, many individuals who manage a limited partnership will establish a corporation, retain sole ownership of that corporation, and have the corporation (which must be adequately capitalized) serve as the general partner. This machination may insulate an individual's personal assets from claims by creditors of the limited partnership, but it is expensive, cumbersome, and adds additional layers of complexity to the business management process because naturally each layer adds legal fees and more documentation—just what every healthcare practitioner in America needs (only kidding)!

Participation in managing the business. Another principal disadvantage of the limited partnership form is that it precludes a segment of its owners, each limited partner, from participating in the management of the business. Although most states provide some "safe harbor" rules specifying those management decisions that limited partners are free to participate in, unanticipated or ambiguous issues are bound to arise, and such rules offer little guidance. What is the penalty for a limited partner whose participation in management affairs exceeds designated limits? The limited partner loses his or her limited liability protection and has personal exposure for the obligations of the business. As a result of this draconian penalty, a limited partner may feel helpless watching the general partner make decision after decision that hurts the business and the limited partner's investment. The prospect of such helplessness often deters prospective investors from committing their capital to a limited partnership.

I once worked with a vascular imaging laboratory that created several limited partnerships for the purpose of purchasing a vascular imaging unit that cost around $49,500 to purchase (ca. 1997). The man in charge of the operation ran a telemarketing unit to solicit limited partnership units, contributing a minimum of about $5,000 (or more) to the endeavor. He closed the capitalization of each limited partnership at $130,000 to cover operating and marketing expenses. Then we had two talented gentlemen who "placed" the equipment into physician offices and trained their medical assistants or nurses to perform limited studies on patients meeting medical criteria for the examination by history or symptoms.

Suddenly, Medicare, which was the primary payor for this type of exam, changed the billing and eligibility criteria to use the equipment and the endeavor was no longer profitable. When the general partners made the decision to fold the partnerships and liquidate the equipment, the investors were very angry and helpless. They could do nothing more than watch as their investment slipped away, the company folded, and the general partners went on to their next endeavors. The payroll and expenses for the general partners were lavish; the scene unbeknownst to me was a repeat of what had happened at another similar corporation that was in trouble with the Securities and Exchange Commission (SEC) for similar questionable activity.

Once the company folded, I received a house call from the SEC, inviting (translation: intimidating and compelling) me to come down and have an informal chat (with a court reporter in attendance and a tape recorder in motion) with a couple of guys (with badges and guns) and to bring all software, notes, and papers in my possession to the little "chat session." I knew very little of what they wanted because I was sequestered from the activity of the telemarketing and its related paperwork and activities, and was only involved in the business and compliance issues on the clinical operations related to the use of the equipment. It left a bitter memory of what can go wrong with limited partnerships and general partners' actions. Add the volatility of healthcare reimbursement rules and technical equipment issues and you have a potential for success or a potential for disaster.

CORPORATIONS

Advantages of Corporations

The corporate form is now considered the leading business form in the United States. If you remember your history classes from high school, Americans saw opportunity during the industrial revolution, where adventurous businessmen and women searched for a way to invest capital without risking their entire personal resources. These risk takers also required a business form that would not require liquidating the business because one or more investors wanted to leave the business. Corporations offered the perfect solution.

Regular (or C) corporations, as they are called, are formed upon filing a certificate of incorporation with the Secretary of State or other designated official in the state where the business will be established. Once established, corporations are recognized by law as distinct legal entities and have the power to act in their own name as persons. Corporations are owned by shareholders but managed by directors, and, upon delegation of authority from directors, by officers as well. These officers and directors carry liability in their decision making and must act in good faith and in a fiduciary manner so as to protect the investors in the same manner and fashion as the other investors in the corporation. Therefore, they need to be insured in the event that their actions are cited in a liability action by one of the shareholders, members, or anyone for that matter. Corporations provide their owners with a number of very attractive features. To name a few, limited personal liability for corporate shareholders/owners as long as certain criteria are preserved, and continuity and transferability of debt or equity interests to other interested parties.

Limited liability for business owners. Corporations are treated by law as separate and distinct entities from their owners. As such, their liabilities are generally treated as separate and distinct from those of their shareholders. Consequently, shareholders of a corporation can generally feel secure that creditors of the corporation will not pursue them to satisfy claims. Also, shareholders are protected from claims by an individual sustaining personal injuries as a result of an action by an employee of the corporation or caused by one of the corporation's products.

Piercing the "corporate veil." Notwithstanding the generally recognized concept of limited liability, important exceptions have developed over the years that permit the "veil of protection" a corporation provides to its shareholders to be pierced so that the shareholders may be personally liable for certain corporate obligations. Although a detailed discussion of this doctrine is beyond the scope of this book, shareholders may be personally liable if

1. They have ignored the formalities required by corporate law, instead treating the business as their alter ego (an example of which would be writing business checks for personal expenses of the shareholders).
2. A shareholder's personal negligence caused personal injury.
3. The corporation is insufficiently capitalized to carry out its operations and the shareholder(s) should have expected its under-capitalization to detrimentally affect a party doing business with the corporation. This is one area in which physicians and other healthcare providers who are unfamiliar with business on a more formal corporate level can run afoul of the system as officers and directors because they are held to the standards of the reasonably prudent, trained business person. Because they do not have the training in finance and corporate operations on this level, they stand at risk as an officer or director.

Notwithstanding these exceptional circumstances, when carefully planned, the personal liability can be limited for the shareholders of a corporation.

Corporate lifetime. State law endows corporations with an indefinite and continuous life. Every document that I have ever seen as a paralegal for many states indicates this very fact on the Articles of Incorporation. This characteristic is attractive to an individual who, investing capital in a business, would not like the business wound up and dissolved on the unilateral decision made by another investor to withdraw capital in the business, or by the heirs in the case of the death of a shareholder. While other business forms can be structured to achieve a measure of continuity, the formality of the corporation does it best. Large corporations survive long beyond the lives and business participation of their initial shareholders and employees. The institutionalization of a corporation can provide significant competitive, financial, and other advantages over businesses that lack such continuity.

Transfer of ownership. Another bonus of corporate participation is that it provides an inexpensive and relatively expeditious means to transfer ownership interests in their business. When there is no agreement between shareholders to the contrary (as may be found in some bylaws), shares of stock in a corporation are generally freely transferable. Indeed, shares of publicly owned corporations, such as United HealthCare, Cigna, and Aetna, are routinely traded on stock exchanges through a phone call to a licensed broker. The convenience of transferability permits investors to move in and out of ownership positions in corporations with lightning speed if their investment philosophy or business circumstances change. By contrast, partners can generally not assign or transfer their interests in a partnership to another party without the consent of the other partners.

When I overlay the personality types that are involved in our healthcare delivery systems, I have seen some long-standing feuds that would make partnership agreements for a huge IPA, PHO, ACO (Accountable Care Organization), or MSO endeavor untenable. The high degree of liquidity is an important advantage of the corporation over other business forms, such as a partnership, where it may be difficult or impossible for a partner to conveniently retrieve money invested in the business.

Unlimited flexibility in structuring management. Another corporate advantage that we see in regular or C corporations is that they offer virtually unlimited flexibility in structuring the management of a business as a result of the distinction made under corporate law between owners (shareholders), managers (directors), and day-to-day operators (officers). Common and preferred stock, preferred and subordinated debt, voting and nonvoting stock, high or low quorum requirements at directors meetings, and veto rights for shareholders and/or directors are only a few of the many tools that owners of corporations can use to customize ownership and management rights and responsibilities through the wordsmithing of its bylaws.

Disadvantages of Corporations

Because corporations are treated as separate and distinct legal entities, they are also treated as separate and distinct taxpayers. Thus, the IRS and most state governments impose a tax on the income earned by a corporation, which must generally file its own tax return and pay its own tax based on its earnings. When the corporation distributes its income or assets to shareholders, another tax is imposed on the shareholders (either an income tax if the distribution takes the form of dividends or a capital gains tax if the distribution results from the sale or disposition of stock). As a result, income earned by a corporation is taxed twice. Despite limited exceptions where the double tax is not a material disadvantage to the shareholders of a corporation, such as the use of crafty income-splitting strategies (which are beyond the scope of this book), the second tax often raises the cost to business owners operating as a corporation. (It sounds so complex! I chuckle when I imagine what it would have been like for Joseph and Mary to have had to discuss tax strategies like this on their way to Bethlehem! But alas, that was then and this is now.)

THE SUBCHAPTER S CORPORATION

The Internal Revenue Code provides a limited solution to the double-level tax problem; it permits certain corporations to elect to be treated in accordance with the provisions of Subchapter S of the code. Unlike C corporations, "S corps" generally do not pay a corporate-level tax on their earnings. Instead, all of an S corporation's earnings flow through to its shareholders who, like proprietors and partners, pay a single level of tax on the income of their business. I have this type of corporation for my consulting firm and, back in the early days, many IPAs I formed chose this form as their structure because of its simplicity.

The pass-through tax treatment of an S corporation is a viable solution to the double-taxation problem. S corporations are not the best choice these days for most IPAs, PHOs, or other networks. The IRS has established specific criteria that limit the eligibility of corporations to elect S status. These limitations are discussed below.

Disadvantages of S Corporations

Limited eligibility. The most significant disadvantage associated with the S corporation form is that the IRS has limited its eligibility to corporations that meet the following four criteria:

1. The corporation must have no more than seventy-five shareholders.
2. All shareholders of the corporation must be United States citizens, resident individuals, estates, or certain defined trusts (corporations, partnerships, and many types of trusts cannot be shareholders).
3. The corporation may not have more than one class of shareholders.
4. The corporation may not have more than a 79 percent interest in any subsidiary corporations.

Any one of these limitations could easily present a problem for many businesses. Moreover, shareholders who initially meet the requirements for electing S status must continue to meet those requirements for as long as this status is desired. Even an inadvertent breach of the qualification rules can lead to termination of S corporation status. For example, if a new business is being capitalized by investors with different investment objectives, it might be desirable to issue two or more classes of stock (for example, common stock and preferred stock). Such is the case in many IPAs developed by primary care physicians, where they want capital and participation from specialists, but as nonvoting members of the corporation. This decision would preclude S status. Similarly, a physician- or other healthcare-provider-owned business that is founded by ten participants may elect S status at the commencement of operations but, after the membership drive and credentialing, the number of physicians or other healthcare provider shareholders may easily exceed seventy-five and thus require termination of S status.

Adverse tax consequences. Aside from the difficulty of qualifying for S corporation status, the S election also can have adverse tax consequences, notwithstanding its pass-through tax treatment. Some of the more important adverse consequences, including limits on the use of debt to create tax basis and prohibition on the use of special allocation of corporate income, must be addressed by appropriately trained, competent professional counsel taking into account your unique circumstances. Make sure the counsel is qualified. I learned this lesson the hard way when my business was audited and a start-up capital transfer from my personal funds was put on the books as cash and I was taxed with interest and penalties by the IRS and the State of Colorado. The accountant who gave me guidance stated that everybody did it that way, and the IRS's response sounded like my mother … (you know, the old "If everybody jumped off the bridge, would you jump too?"). The IRS stuck to its position and I paid … lots! CPAs are just like other professionals and craftsmen—some good, some not as good. Caveat emptor!

DOUBLE-TAXATION CONSIDERATIONS

There is an alternative for businesses that may not qualify for S corporation status yet seek the advantages of incorporation (limited liability, transferability of interests, management flexibility, etc.) by maneuvering around the double-taxation issues.

Income earned by regular corporations is taxed twice: first on the business-entity level and later its shareholders are taxed again on the distributions they receive from the corporation. If, however, the corporation has no income, it pays no tax. Therefore, sophisticated entrepreneurs often operate their corporations in a manner designed to limit (not maximize!) income. This is especially true in the IPA or PHO environment where medicine is being practiced and the corporation should he kept (in my humble opinion) cash-poor to thwart any attempts at searching for deep pockets and finding them at the IPA/PHO level.

Here, one would want to keep the deep pockets at the MSO level and not have any clinical activity going on at the MSO level whatsoever. The shareholders' maneuvering to limit corporate income, yet put as much money in their pockets as possible, must be carried out carefully and delicately. When the corporation can justifiably pay all of its income out in the form of tax-deductible compensation to its employees and operating expenses, the business has no income left on which to pay tax. Another common means to limit (or even eliminate) corporate income is to finance the business with debt provided by shareholders. The corporation's payment of interest on the debt is also deductible, and the repayment of the loan principal amount is tax free. This compares favorably with the distribution of equity to the same shareholders, which could be subject to a capital gains or ordinary income tax. This is where many err in the start-up of S corporations. GET GOOD ADVICE from your tax advisor!

Attorneys and accountants often seek additional loopholes to avoid the double level of tax imposed on income earned by corporations. Remember that you can avoid paying taxes, you just cannot evade them. Many of these techniques are useful but there are limits to such creativity, and the IRS acts to close such loopholes by constantly developing and refining its volumes of Commerce Clearing House directories of restrictions. As a result, even with superb tax planning, most successful corporations ultimately pay a corporate-level income tax, which can be substantial. Many businesses grudgingly find themselves paying the additional tax as the price for the benefits of operating in corporate form. If your long-term goal is to keep the clinical model poor and put the cash in the MSO, get good counsel familiar with healthcare issues and the organizations.

THE LIMITED LIABILITY CORPORATION (LLC)

The LLC offers the following tax incentives and for this reason has been most popularly chosen over the other available forms for IPA/PHO development throughout the nation, namely

- Limited liability (to protect its owners from becoming personally liable for the debts of the business)
- Pass-through taxation (to avoid the double taxation attributable to regular corporations), no restrictions on permitted owners (to eliminate the burdensome numerical and other requirements established for S corporations)
- No restrictions on active participation (to ensure that, unlike limited partnerships, all owners could be active in managing the business without jeopardizing their limited liability protection)
- Operational flexibility (to let owners structure the management in a way that satisfies the concerns and requirements for each business)

As a consultant and a paralegal, I carry a directory of statutory citations with me when I go out to client sites to work with start-up groups and usually can download the actual statutes from a source such as Lexis®. This enables clients to read through the statutes so that they can ask intelligent questions of their attorney, accountant, and of their consultants.

In selecting any professional consultant expertise, these things should be made available to you so that you do not have to spend the time searching on your own. You pay us to shorten the research and learning curve. In this chapter, I have exhausted my novice level of competence in explaining the merits of the myriad corporate forms available. If you need help with corporate formation, this chapter is here to give you high-level information to better formulate your questions with competent local counsel that can help you evaluate your particular situation, local regulations, and tax laws.

7 The Steering Committee Gets Busy

Step-by-Step Instructions for What to Do and How to Do It

STEERING COMMITTEE TASK LIST

Background/Understanding of Task

Develop a Statement of the Committee's Purpose

[Below is an example of the layout and the items that should be covered in the background section.]

Twenty urologists associated with this single-specialty IPA would like to move forward and develop an independent provider organization for single-signature managed-care contracting in cooperation with their affiliated hospitals and ancillary providers.

As more fully identified below, the scope and purpose of this project is to design and assist in the development of a Provider Organization (PO) and a Management Services Organization (MSO), to assist the physicians and other healthcare providers affiliated with the group in and around the Metropolitan Statistical Area (MSA) of Dupage County, Illinois and the Greater Chicagoland area so as to prepare for the demands of managed care.

To the extent legally permissible in accordance with the applicable regulatory constraints, the design of the PO will provide the physicians involved and committed to the project the opportunity to bid and hopefully participate in managed-care arrangements while maintaining a completely independent practice. Each physician or group practice shall continue to bear the responsibility for the cost of all assets, equipment, and supplies of their existing practices. The physicians shall not be relieved of any management and administrative responsibilities associated with their respective practices, except as may be required by then-existing regulatory requirements.

It is understood by both the physicians and consultants that the PO shall be designed to provide a structure for facilitating cooperation among the independent individual practices for the purpose of providing high-quality, cost-effective urologic care in a managed-care setting. The role of the PO shall be as an Independent Practice Association (IPA), and it shall assist those affiliated physicians with opportunities to develop new and unique healthcare products, while facilitating a sharing of the financial risks and responsibilities associated with the same.

To this end, the PO shall develop and provide a consortium of orthopedic surgeons for centralized utilization management (UM), quality improvement (QI) and credentialing, and data collection services for the entire network of providers. The PO generally is not intended to, however, become involved in the day-to-day operations, business risks, management, or ownership of any of the individual practices that are part of the PO, except as may be required by then-existing regulatory requirements. This is not to prohibit a change of vision for the future, which may eventually combine the practices into one corporation, if desired by the parties.

At first, the PO will provide a structure by which providers may be assisted and supported with the development of managed-care arrangements. Eventually, the design shall be enhanced so that the PO is sufficiently integrated with all affiliated practices so as to take on financial risks associated with capitation arrangements.

Long-range plans include the development of a second organization commonly referred to as a Management Services Organization (MSO), which, if developed, shall be designed and developed to provide all the support and management services associated with the PO. The MSO shall provide expertise in centralized management of the business aspects of all the practices in the PO.

The MSO may assist with hiring all personnel, provide a mechanism for group purchasing, contracting, marketing, data integration, performing utilization and quality monitoring, billing and collection services, among other functions such as the group may desire, all to the extent legally permissible by then-applicable regulatory requirements.

Remember that this need not be a masterpiece, but it must be a functional road map and communication tool for the attorney, the accountant, and to refocus the steering committee if it gets lost along the way.

Approach

Together with a professional consultant (as needed) with expertise in the management and development of such networks, the steering committee then needs to work on the following tasks:

1. Development of shared vision
2. Organizational development of the IPA
3. Organizational development of the MSO
4. Market focus
5. Business plan

Each phase should conclude with a go/no-go decision and a deliverable. These deliverables should be spelled out for the consultant by the steering committee so that you can gauge the effectiveness of the consultant and the expenditure of the capital dollars when reporting progress to other interested parties.

Development of the Shared Vision

The purpose of the work session would be to develop a clear statement of goals:

- Develop a clear understanding of single-signature contracting.
- Develop a consensus on managed-care direction:
    1. Right-sized network. (This takes market research that the committee may delegate to a marketing subcommittee or may purchase information from a company such as Sachs in Chicago.)
    2. All or selected providers participate (steering committee decides what will be logistically feasible to get the work done).
    3. Capitation goals and objectives must be visualized and strategized.
- Identification of physician interests in leadership to serve the group as developmental leader for the following tasks:
    1. Creation and monitoring of bylaws
    2. Credentialing and membership
    3. Utilization management
    4. Quality improvement—standards of care
    5. Medical directorship
    6. Marketing
    7. Risk management
    8. Finance and budget
    9. Business development
    10. Information systems
- Develop a statement of target market intent.

Often, the consultant assists the team in designing, refining, and then documenting the vision. When working the project as the consultant, I have found that working with each physician leader to outline work tasks to be completed by the physicians and the consultants together works the best. In this way, the work is not overwhelming to anyone unfamiliar with the design and development tasks. From time to time, the steering committee may need to be augmented by temporary assistance from outside sources on an independent contractor basis for secretarial tasks, mailing, and telephone calls to pull projects together.

From the outcomes of these activities, the steering committee and the consultants usually develop statements of objectives, vision, and mission. These statements are then returned to the shareholders for review before final ratification by the membership.

Organizational Development of the IPA

The steering committee must next either delegate the following tasks or carry them out themselves:

1. Select an experienced, health law attorney.
2. Select an accountant familiar with healthcare and capitation issues.

3. Organize and schedule a meeting of interested fellow participants who will become the general membership participants in an organizational goals workshop.
4. Create lists of all the IPAs, PHOs, PPOs, and HMOs in which participating physicians are members.
5. Create a list of available resources for utilization review and quality control.

Because the business plan will involve the sharing of sensitive market data that might border on antitrust activity, the steering committee must look to the selected attorney for guidance in developing the organizational papers that permit business planning and work the leadership team through the major decisions.

Additionally, the steering committee must also look to the selected attorney for compliance with all then-current regulatory issues and limitations set forth by Stark I and II, antitrust, fraud and abuse laws, and regulations for the chosen organizational structure, and both to the attorney and accountant for consideration of any IRS and tax status implications, and for the requirement of any local regulatory concerns.

Next, the steering committee needs to design and direct work tasks to achieve the following:

1. Develop a statement of purpose to help participants agree on the goals and objectives of the organization. The proposal may, of course, include goals that will not be immediately realized.
2. Prioritize activities that the PO will pursue and determine how the PO will operate. The scope of the services is dependent on the defined goals and objectives of the organization.
3. Develop a business plan for both initial and subsequent operations of the organization.
4. Develop the level of understanding for the different task committee leaders.

After these four steps, the steering committee needs to work together with the selected consultant team to develop the initial documents necessary to form the organization, including

1. Articles of incorporation or organization; bylaws
2. Shareholder agreements
3. Participating physician agreements
4. Participating payor agreements
5. Participating hospital agreements
6. Participating ancillary agreements
7. Utilization management policy documents
8. Quality assurance/improvement policy documents
9. Marketing plan
10. Information systems plan
11. High-level pro forma budget
12. Credentialing criteria and procedure

13. Grievance procedures
14. Reporting requirements and their schedules

Having guided many organizations through this process, as you can deduce from what I have shared, there is a lot that goes into preparing these documents to get a rough draft ready for attorney review. Working together with several IPA/PHO (Physician Hospital Organization) steering committees and executive boards over the years, you can see that my methods have been honed to a science. Most of these documents should be semi-prepared by your consultant team, and you should not be paying for document development from the blank paper stage. What you pay for in consulting expertise is to go beyond document preparation, to shorten the learning curves, and to share lessons learned.

If you are speaking to a consultant or an attorney who has never developed these documents before and has to charge you to start from scratch, keep looking. Consultants who have experience in this particular area of healthcare are few and far between, but they are out there.

Organizational Development of the MSO Required of the Steering Committee

The organization development of the MSO required by the steering committee involves

1. Selection of the health law attorney
2. Selection of the healthcare accountant familiar with capitation and physician reimbursement
3. Development of an SEC memorandum for investment by other members if the capitalization will be more than $1 million
4. Participants' decision of physical location of business office

Because the business plan of the MSO will involve sharing of sensitive market data that might border on antitrust activity, the team must look to the selected attorney for guidance on developing the organizational papers that permit business planning and work the leadership team through the major decisions:

- User fees—revenue allocation
- Departmental objectives
- Equity decisions
- Selection of staff and equipment
- Ancillary contracts
- Ancillary equity participation

In no event should the steering committee proceed with the development of any business plan based upon any shared market data of the members unless and until clearly identified safeguards can be established to the satisfaction of the group or its legal counsel. With the members on a single-signature corporation with common bottom lines and the intention of a single-signature capitated contracting, it should be possible with appropriate legal guidance to pool market information with less restriction and risk by the formation of an economically integrated MSO.

Market Focus

These steps come next on the steering committee's list:

1. Provide hospital partnership goal, if applicable.
2. Provide capacity of each geographical area for new and converted business.
3. Provide managed-care contracting parameters (if and when attorney states that this is acceptable).
4. Provide current managed-care activity by volume of claims and dollar value.
5. Facilitate request to physician offices for collection of claims data for outcomes measurement in accordance with HEDIS, Leapfrog, etc. guidelines.

The objectives of the market data research are as follows:

- Define the breadth of geographic coverage intended in the contracting effort.
- Identify the competition.
- Identify market sensitivity.
- Identify needed market services.

Development of a Business Plan

Building on the previous phases, the necessary consultants would work closely with the steering committee to build business strategy into a formal written business plan that includes the following components:

- Description of the business
- Description of the products
- Description of the market
- Description of the market size in terms of incidence frequency rate of disease, surgery, injury, (need) for the specialty(ies) in question. For example, heart surgery, age 44–65 = 48:10,000 population, etc.
- Market niche
- Description of coverage
- Marketing strategy
- Financial strategy
- Administrative/management structure
- Quality improvement/monitoring of outcomes
- Resource requirements
- Timeline for development and implementation

Where to begin? When faced with the decision to do something, the first activity necessary is to select and organize a steering or investigative committee that will act as the temporary board and incorporators of the entity.

This chapter addresses the organization of the steering committee, its tasks, and areas of consideration. Although it is not exhaustive, it will cite the more crucial areas that the team faces in getting the organization off the ground.

ORGANIZING THE STEERING COMMITTEE

Steering committees should have several members, each with different expertise or interests. I address several types below, including the multi-specialty IPA (Independent Practice Association) or PHO (Physician Hospital Organization), the single-specialty IPA or PHO, and the non-physician IPA.

MULTI-SPECIALTY IPA OR PHO

This design, being the largest and most diverse team, should be made up from different factions of the medical staff if you plan to have an all-inclusive, successful, risk-bearing organization. Whether you are designing a multi-specialty IPA or PHO, this committee should be composed of primary care, specialty, and perhaps even a sub-category of hospital-based physicians. The needs in each category are very different. The team I have found most successful in the design of an IPA steering committee is to have one of each of the following: family practice, pediatrics, ob/gyn, general internal medicine, gerontology, orthopedics, cardiology, rheumatology or physiatry, general surgery, gastroenterology, diabetology or endocrinology, neurology, and psychiatry, as well as radiology, pathology, emergency medicine, and anesthesia. If I have a team such as this, I can build a Med-Surg IPA, a Medicare product, a Medicaid product, a Workers' Compensation product, and a product to service motor vehicle and casualty insurance products for both PPOs (Preferred Provider Organizations) and exclusive provider organizations. In this way, my organization will stand ready to service all lines of insurance and third-party payors available in the marketplace. I also have a well-rounded team that complements one another in clinical areas. If I can find the members of the staff with the qualities and specialties to make it happen, my steering committee has seventeen dedicated people who will need to serve on at least one subcommittee.

Some consultants might argue that this group is too large to get any work done, but I disagree. I look for special qualities in a steering committee: specifically leadership and communication skills; knowledge of the marketplace; a basic understanding of capitation, and a respect for risk rather than a fear of it; a competitive spirit; and an acceptance that this product is needed and desired by both the community and the physicians in that community. I also make sure that each has a good sense of lair politics or charisma to persuade others to roll up their sleeves and join in the work party as needed, and most of all, a genuine interest in taking control of medicine and putting it back in the hands of physicians where it belongs. Oh, and one more thing: a sense of humor! When all these qualities are present, they are educated as to what the job entails and they come to meetings to be productive, rather than to sit and commiserate. The team I try to develop for this position is too busy to waste time commiserating! Steering committee meetings should have an agenda and stick to it. Parliamentary procedure is a must; therefore, a pro-tem board should he nominated and those not elected as officers should be members-at-large. This group should be prepared by a consultant team of an experienced healthcare organizational development consultant, a CPA (Certified Public Accountant) familiar with capitation, an experienced health law or transactional law attorney, and a healthcare actuary. From time to time, I like to add a few members of the business community to lend their ideas to the group.

Single-Specialty IPA or PHO

Unless you have some very specialized niche, single-specialty groups are not for every market, except for certain specialties that deliver services based on a rider inclusion for those services to be covered by contract, such as chiropractic care, podiatry, optometry, or for special services that may be either esoteric or specialized high-cost services that might sustain themselves more easily through a capitated carve-out from a generalized medical loss ratio budget. These teams include, perhaps, hospital-based subspecialties such as neonatal or prenatal intensivists, mental health and substance abuse services, anesthesia, radiology, pathology, emergency medicine, or other specialties such as high-risk obstetrics, or limited-practice groups that specialize in a particular area of medicine as consultants with a short-term relationship to the patient and no real ongoing follow-up once the crises are over, as is seen in the routine practice of office-based medicine.

It has been my experience that many managed-care plans would rather spend the time and effort contracting with an entire continuum of care rather than one specialty focus group, if given a choice. The reason for this is that the yield for the time spent is greater with the full-service group. Now that is not to say that there is no validity in single-specialty groups, but rather that one might experience a more difficult time of it when attempting to attract the attention of a payor entity that is focused on building a network quickly and inexpensively.

In working with a group of orthopedic surgeons who wanted to build an orthopedic IPA only, we realized quickly that in order to monitor utilization, control costs, and quality concurrently, the physicians would have to hand-pick a team of support players. For instance, who was going to put the patient under general anesthesia or manage monitored anesthesia control (MAC) while they operated? Who would carry out physical therapy orders? Who would dispense prescription medications to their patients? Who would manage any medical comorbidities?

In a capitated carve-out situation for orthopedics, in order to gain more clout and more control of the premium dollar, they would have to subcontract with all these players on the cursory list above. This is something to consider as you develop your goals and objectives for development of your group, because the management of the entire continuum of services related to orthopedic cases is not only the way to gain data and truly control utilization, but it is also a value-added service to the payor that centralizes the group's activities rather than providing still another fragmented piece of the delivery continuum. Payors like that and are usually willing to transfer dollars and delegate responsibilities to the group well developed for that intent.

Management Services Organization (MSO)

The steering committee for the MSO has a similar task to that of the IPA and PHO in its development, with the exception that it must have leadership at the helm and an ability to speak at many levels of understanding. These people are the linguistic translators. They can speak medicalese, legalese, accountingese, actuarialese, insurance-ese, and computerese, and they can communicate to the rest of us and tell us what they want the organization to do for them through departmental development and delegation of responsibilities. Most often, the steering committee for this group sets out to steal a contracting expert from a local managed-care organization, thinking that this person will be a sort of "Messiah" who will handle every issue from the moment that he or she is hired. NOT!

The first thing that the MSO steering committee must do is decide upon the activities and role that the MSO will play in the management of the clinical team that is related at arm's length, and who may be the same investors in a different organization with a different corporate setup.

When the steering committee begins its due diligence and strategic planning, one of the first steps it must take is to decide on a corporate form. Some of the more popular and possible appropriate forms the corporation may take (whether an IPA, PHO, or MSO) are general partnerships, limited partnerships, regular or C corporations, Subchapter S corporations, and the most popular (and the new darling of the bunch), the limited liability corporation. In the following pages, I compare those listed above. Naturally, it is imperative that you seek appropriate professional competent counsel for your specific needs. The comparisons set forth are very general as this book is intended for a readership from a variety of locations and corporate intents.

Other Concerns of the Steering Committee

Thus far we have discussed the different constituencies and corporate forms available to provider networks. What else does the steering committee have to do? Let's talk a little about strategic planning and business plan development. It is likely that the steering committee will have had the most in educational opportunity and time for consideration, as well as more time to develop leadership and communication skills than the other late-comers for the development of your network. Therefore, while the steering committee may or may not become the executive board, it may well become, by default, the business development committee. This is especially true in smaller groups.

Following is a checklist that I use when meeting with steering committees to give some structure and an agenda to the meeting. Although most have capital, they are by no means in a position to waste time or money unnecessarily with a consultant with minimal experience and a "we will both learn it as we go along (since I am billing you for my time)" attitude. I have been called to do clean-up detail on some botched jobs on those who carry briefcases and business cards where some unsuspecting client will become their "n = 1" network project. I now know how some surgeons feel in the surgical suite. Professional decorum makes you keep your mouth shut and do your best. I want this hook to be your personal "fishing manual." I am of the philosophy that if I teach you to fish here, you can eat for a lifetime on your own skills and only use consultants for that which you cannot do for yourself, or that which you need to shorten your learning curve to save time and money with a private lesson.

One way that you can put this information to good use is to read the next chapter, which focuses on the tasks performed by the steering committee, writing up the outline of the business according to the layout and template I have supplied, and then using it to guide discussions with your chosen accountant, consultant, and attorney so that they can back into the best organizational structure for your group. It is an exercise for which you really do not need handholding, but you do need privacy to speak openly between the elected or appointed members of the steering committee to determine, without influence and interference, where your key concerns, goals, and objectives are for the group.

Expect the exercise and write-up to take about six weeks to complete if you only meet for an hour or two at each session. In past organization steering committee meetings, I have done this in a two-day marathon session with ten to sixteen steering committee members present in about sixteen hours with working lunches, white-boards, and flipcharts in a rented meeting room. At the end of the two days, the report was drafted and ready for grammar checks and final tweaking. The following weekend, we reconvened with three hours set aside for attorney discussion of the options, and then two more hours with the accountant. All questions were asked and answered. One thing I learned early on: send the attorney and the accountant the draft form of the document and have them review it before their arrival. After the five hours with the attorney and accountant, we adjourned for lunch and the private working lunch meeting was held, decisions were made, and a strategy and implementation plan was ready for publication.

Another way you can work this is to take the following chapter and turn it into a "fill in the blanks" document. As you make decisions, complete each section with the particulars. You will not spend as much time during the meeting to write up text from scratch.

8 Guidance for the Utilization Management and Quality Improvement Steering Committees

Utilization management and quality improvement committees should be developing, from the very start, a knowledge of how payors think. First, remember these three concepts:

1. Utilization equals expense instead of revenue.
2. Patient satisfaction and clinical outcomes are the keys to renewed contracts.
3. The appropriate care at the appropriate level at the appropriate place at the appropriate time by the appropriate provider is paramount to success.

An insured's benefit equals expense and risk exposure to the payor's bottom line. Therefore, for your organization to represent benefit and attractiveness to a payor, you must be able to manage utilization and also show your management plan on paper. In addition to a growing set of standards and measures to accomplish this, the information systems available to support measurement have improved. As a committee, your task is to determine what will be measured, how you will obtain the data, what tools and reference standards you will need to get the job done, and then identify candidates from an applicant pool of potential employees or outsourced companies that can handle the project with your oversight.

Myriad administrative or transaction data—medical and pharmacy claims and laboratory results—have increased in both availability and comprehensiveness, thus providing a rich and convenient information source from which organizations can evaluate healthcare. This could also present information overload if you are not selective in data gathering. Equally important are the tools and technology that encode standards of care to provide an efficient and robust way to assess care compliance with these standards. Many are out there and more continue to be developed and refined. Take the time to consider the "Milliman Healthcare Management Guidelines" because many health insurers rely upon these guidelines as a basis of payment and utilization and payment decision making. Other guidelines are also available and widely adopted in addition to Milliman, including evidence-based medicine systems from Johns Hopkins, Elsevier, Symmetry EBM Connect® from Ingenix, and others.

You will use these tools to

- Identify disease and care opportunities, including "gaps" in care for patients and populations.
- Identify both high-performing physicians and areas that need improved physician compliance with prescribed care.
- Identify diagnostic tests or treatments that are unnecessary or potentially harmful, with the ability to determine the pervasiveness of these tests in your assigned populations.
- Identify patients with indications of poor disease control, such as low adherence to prescribed medication regimens.
- Reduce potentially harmful drug-to-drug or drug-to-disease interactions.

FIRST THINGS FIRST

To start an organization and make sense of your provider service agreements, you must have at least the framework of a utilization management policy on paper. Following is an outline of the elements that must go into these documents:

Utilization Management Program Outline

I. Statement of Purpose (Here is a model list to get you started.)
   A. Identify disease and care opportunities, including "gaps" in care for patients and populations.
   B. Identify both high-performing physicians and areas that need improved physician compliance with prescribed care.
   C. Identify diagnostic tests or treatments that are unnecessary or potentially harmful, with the ability to determine the pervasiveness of these tests in your assigned populations.
   D. Identify patients with indications of poor disease control, such as low adherence to prescribed medication regimens.
   E. Reduce potentially harmful drug-to-drug or drug-to-disease interactions.
II. Objectives
   A. Monitor medical needs of each patient
   B. Monitor the level of care
   C. Ensure appropriate resource management
   D. Develop and evaluate normative data
   E. Analyze global payor trends
   F. Study statistical data and use information to guide the organization
   G. Monitor effect of policy change
   H. Recommend further policy development as required
III. Participants
   A. Board of directors
   B. Medical director

  C. Members
  D. Support staff
  E. UM committee
  F. Tenure of each position
IV. Confidentiality Policy
V. Conflict of Interest Policy
VI. Plan Reappraisal
VII. Activities to Be Studied and Frequency of Each Study

The quality assurance and quality improvement committee must also be prepared. Its own outline for a program should be developed along the lines of the following outline:

Quality Improvement and Assurance Program

I. Statement of Purpose
  A. Monitor and evaluate quality of care
  B. Pursue opportunities for improvement
II. Plan Activities
  A. Pre-contract on-site review:
    1. Review tool to be used
    2. Summary and action plan relative to findings
  B. Patient complaint procedure:
    1. Objective
    2. Procedure
    3. Findings and related action plans
  C. High risk reviews:
    1. Objective
    2. Procedure
    3. Summary and action plan relative to findings
  D. Adverse outcome review:
    1. Objective
    2. Procedure
    3. Summary and action plan relative to findings
  E. Medical records management and documentation quality:
    1. Participating provider documentation review
    2. Procedure
    3. Evaluation tool
    4. Summary and action plan relative to findings
III. Participants
  A. Board of directors
  B. Medical director
  C. Members
  D. Support staff
  E. UM (Utilization Management) committee
  F. Tenure of each position

IV. Confidentiality Policy
V. Conflict of Interest Policy
VI. Plan Reappraisal

Activities to Study and Frequency of Each Study

Keep in mind that you will most likely be asked the following questions by the health plans that deal with your new organization:

Frequently Asked Questions by the Health Plans (Don't Be Caught Without An Answer!)

- Has the National Committee of Quality Assurance (NCQA), American Association of Preferred Providers, or other external organization reviewed your organization? Has a self-assessment been conducted? If yes, please provide a copy of the results of the most recent assessment.
- Please give a brief description of your current quality improvement initiatives, including the method used, data sources, and any changes in the outcome that have resulted.
- Please describe in detail your credentialing process for providers and facilities. Do you routinely obtain peer evaluation as part of the credentialing and recredentialing process? If so, describe the process.
- What quality-of-care services measures are used?
- For credentialing purposes, how do you evaluate pending malpractice cases and physicians undergoing substance abuse treatment? What follow-up procedures are in place?
- Is there a functioning Utilization Management/Quality Assurance Committee? How often do they meet?
- Please attach a list of the names and specialties of the physicians who are on the Network's Utilization Management/Quality Assurance Committee(s).
- What are the responsibilities of the Utilization Management/Quality Assurance Committee?
- If the Utilization Management Committee is a separate committee, please address each individually.
- What criteria are used by the Utilization Review and Quality Assurance Committees to assess the quality of the care provided by the organizations with which the network has contracted?
- What reports does your administration system produce to allow the Utilization Review and Quality Assurance Committees to evaluate performance?
- Indicate the specific role of outcome measures in the delivery of care within your group. Provide samples of the instruments used to collect patient-reported outcomes information.
- Indicate the specific clinical areas in which outcomes have been or are currently being assessed, and provide results (whatever available) for each area.
- Describe the benefits that have been derived from the systematic assessment of outcomes.

- If condition or procedure-specific outcome studies have not been completed, please provide the rationale and outline any pertinent future plans.
- Describe any constraints concerning the systematic study of patient outcomes, which health plan should be mindful of an outline, suggested approaches to addressing these constraints.
- Identify the clinical areas in which your group would be prepared to implement outcome studies at the outset of the proposed program.
- Indicate the physician-specific quality indicators currently tracked within your group.
- Describe the method(s) by which the data are adjusted for severity of illness or otherwise risk-stratified.
- Provide the summary quality indicator reports for the past two calendar years.
- Describe the specific process (and provide example) that promotes the continuous improvement in healthcare delivery as evidenced by improvement in quality indicators over time.
- Indicate the approach used to systematically analyze and compare clinical practice, procedure results, and quality indicators. If the data are not tracked, please provide the rationale and outline any pertinent future plans.
- Describe the circumstances (protecting individual identity) surrounding any physician who had his or her privileges limited or revoked, or any group staff member who was disciplined or dismissed as a result of the QM program.
- Describe three examples that demonstrate the effectiveness of the QM program, outlining the underlying issue, the method used to identify and quantify it, the specific intervention, and the reassessment process.
- Describe the methods currently used to assess patient satisfaction with the services provided by your group and provide a sample of the instrument.
- Provide a summary of the results of the last two patient satisfaction surveys.

Documentation Quality

The medical records documentation guidelines for your network organization should probably mirror the NCQA standards. While electronic medical records systems generally force many of these points, there may be physicians who have not yet adopted an EMR (Electronic Medical Records) system or who have adopted one that does not quite do the job necessary.

The evaluation tool used most often by many health insurers includes the following questions:

- Do all pages contain patient ID?
- Is there biographical/personal data?
- Is the provider identified on each entry?
- Are all entries dated?
- Is the record legible?
- Is there a completed problem list?
- Are allergies and adverse reactions to medications prominently displayed?

- Is there an appropriate past medical history in the record?
- Is there a pertinent history and physical exam?
- Are lab and other studies ordered as appropriate?
- Are working diagnoses consistent with findings?
- Are plans of action/treatment consistent with diagnoses?
- Is there a date for a return visit or other follow-up plan for each encounter?
- Are problems from previous visits addressed?
- Is there evidence of appropriate use of consultants?
- Is there evidence of continuity and coordination of care between primary and specialty care physicians?
- Do consultant summaries, lab, and imaging study results reflect primary care physician review?
- Does the care appear to be medically appropriate?
- Is there a completed immunization record?
- Are preventive services appropriately used?

Adverse Outcome Review

Another task you may endeavor as a committee is to select certain criteria to monitor for outcomes that, in some instances, may have never needed to progress to such a high level of acuity had the appropriate intervention been initiated at the appropriate time, namely

- Cancer of the breast (female)
- Cancer of the cervix
- Cellulitis
- Cancer of the colon or rectum
- Diabetic coma/ketoacidosis
- Gangrene (angiosclerotic)
- Hemorrhage secondary to anticoagulation
- Hypertensive crisis
- Malignant hyperthermia
- Hypokalemia due to a diuretic
- Mal-union/non-union of a fracture
- Perforated or hemorrhaging ulcer
- Pregnancy-induced hypertension
- Admitting diagnosis of pulmonary embolism
- Readmission within fourteen days for same diagnosis
- Ruptured appendix
- Admitting diagnosis of septicemia status asthmaticus
- Dehydration of a child under age two
- Low birth weight/prematurity less than 2,500 grams
- Urinary tract infection

Add to this list the entire range of CMS' (Centers for Medicare & Medicaid Services) "Never Events," for which most insurers now include reasons for nonpayment in their contract.

With the development of both UM and QI programs, be sure to add in some language about due process and arbitration or mediation in the event that someone who was the subject of the review disagrees with the outcome and findings and perhaps the decisions made. The NCQA requires that there be a grievance mechanism for all accredited health plans; therefore, your network should also mirror this policy. You will be well on your way to establishing the network as one who is serious about managing quality and utilization while rendering appropriate patient care. Now that you have seen the outlines, go to work. Remember that I expect you to make a protocol that is better for me as a patient than what my insurance company would design. After all, you folks are the trained healthcare practitioners, and you know quality or lack thereof when you see it!

Oh, and by the way, your work is not done when you finish the laundry list in this chapter; you have to work with the finance committee members on their tasks in the next chapter in order to achieve real integration and alignment.

9 Network Financial Management

The Intersection of Finance, Utilization Management and Capitated Risk Management

"Ugh. Yuck. Finance! If I wanted to do business, I would have gone to business school."

—Anonymous physician

Well, ladies and gentlemen, you volunteered for this steering committee, so let's get busy.

Because an IPA (Independent Practice Association), PHO (Physician Hospital Organization), or MSO (Management Services Organization) usually takes capitated risk, this chapter is designed more as a reference and planning tool for the finance committee, and the utilization and quality committees. The task is to learn to understand how the numbers work and how to supervise your administrators to meet the goals and objectives of the organization. Nonphysician providers may find many of the reports useful as well to stimulate reports that mirror these but that are more germane to their areas of specialization. These reports will also provide a great start on outcomes reporting for internal management and marketing.

Let's start with the basics. Often I find that medical practices, especially primary care practices that are already assuming capitated risk, do not close out their month-end reports as they should. Therefore, forgive the elementary level of this early part of the chapter, but I have learned the hard way to assume nothing. I also have been made painfully aware that in the PHO setting, if the billing and financial management team is "borrowed" from the hospital side, often the physician and other non-hospital billing issues are foreign to most hospital billing experts, and therefore they find it difficult to truly manage capitation. This last point, although unintentional, often contributes significantly to misalignment and disintegration of PHOs.

For these reports, I have blended my experience as a practice manager, hospital business office coordinator, and my HMO (Health Maintenance Organization) reporting capability background from my stint in provider relations of a large HMO capitated plan. These are the reports I have found most helpful. Most every one can be achieved using Microsoft Access® or another relational database software tied to a centralized data depository at an MSO or similar venue.

FINANCE REPORTS

- *Monthly and annual aged trial balances.* A network-wide report demonstrating the accounts receivable (A/R) status in the following format:
  - Patient's last name, first name, middle initial, responsible party
  - Current balance at 0–30 days, 31–60 days, 61–90 days, 91–120 days, 120–150 days, and 151–180 days
- Employer name, Payor ID, Subscriber ID, Group ID, Home phone, Work phone
- *Monthly, quarterly, and annual adjustments to open balances.* This is usually a report detailing all A/R adjustments made as debit adjustments.
- *Monthly-quarterly insurance receivables grouped by payor.* Current–30 days, 31–60 days, 61–90 days, 91–120 days, 121–150 days, 151–180 days, and 181–120 days for all fee-for-service reimbursements expected. Listed by patient and date of service.
- *Monthly and quarterly collections status report.* A report showing what has been done to collect receivables due, pended claims, suspended claims, notations on the account, who has worked on the account, and the present status and/or recommendations.
- *Monthly, quarterly, and annual paid-to-billed ratios (discounted fee for service claims).* A report showing the present-day status of all reimbursement for all payors using discounted fee-for-service methods.
- *Monthly collections activities by billing specialist monthly and quarterly suspended claims reports.*
- *Monthly, quarterly, and annual paid-to-billed ratios—capitation.* This is a report showing the ratio of revenue collected through capitation, copayments, fee for service for noncovered services of an elective or cosmetic nature, against what would have been collected using a standard conversion rate for productivity-based reimbursement on an encounter-by-encounter-only basis.

Next, create a report that lists data for each individual capitatated providers in the following format:

Provider Name	MM/YY	Health Plan Name	Paid-to-Billed Ratio

Table 9.1 may be of help for those of you unfamiliar with this method of monitoring your A/R and revenue cycle.

UTILIZATION MANAGEMENT REPORTS

- *Monthly and quarterly procedure rate frequency for all primary care providers by practice and by individual physician.* Analysis by physician of all evaluation and management codes used in encounters and billings, regardless of reimbursement type.
- *Monthly and quarterly procedure rate frequencies by special procedures by PCP and by specialty (IM, FP, Peds).* Analysis of procedural frequencies

TABLE 9.1
Paid-to-Billed Analysis

In order to derive your paid-to-billed ratios, you must calculate the following information:

ABC Health Plan

Step One:

$34,000	Capitation revenue paid
$240	Copayments and deductibles paid
$1,305	Elective and noncovered services paid by patients (paid, not charged, and in A/R!)
$35,545	Total health plan derivative revenue

Step Two:

$35,545	Total health plan derivative revenue
$26,405	Actual billed services at usual and customary charges

Step Three:

134.6%	Paid-to-billed ratio

Note that good capitated performance is in the 130 percent to 160 percent paid-to-billed range.

for certain in-office procedures, for example, otic lavage, EKG, spirometry, pap smears, biopsies, etc.

- *Quarterly procedure rate frequencies by referral specialists/others.* Analysis of evaluation and management codes used for each physician, grouped by specialty.
- *Quarterly procedure rate frequencies by referral/specialists for procedures ordinarily performed by PCPs.* Analysis of procedural frequencies for certain in-office procedures, that includes: otic lavage, EKG, spirometry, pap smears, biopsies, etc.
- *Month-to-date and year-to-date utilization by provider.* A report showing each physician's productivity on a monthly basis by CPT code reported.
- *Quarterly formulary compliance by PCP.* A report showing formulary compliance rates by network and referral physicians, including out-of-network physicians. This report should show only individual performance. Layout should be by drug category, for example, antihistamines, antibiotics, analgesics, etc. Reports should also contain peer comparisons.

  Further detail would be helpful showing any patients who are in stop loss or catastrophic categories, listing patient name, date of birth, identification number, medications utilized, and diagnoses reported from claims experience as a trailer report. This report would be helpful for the physicians and case management team to be able to closely examine and reevaluate patients on complex polypharmacy with concomitant disorders.
- *Quarterly formulary compliance by referral physician.* This report should include all specialists in all classifications who may prescribe drugs to patients for covered conditions for which an active referral is on file from a network physician. Layout should be by physician, by drug category, and

should include utilization data inclusive of any drugs prescribed that were noncompliant with the established formulary, and drugs that are classified as potentially habit-forming or highly controlled.

- *UM/QI Reports:*
  - Monthly and quarterly laboratory/pathology specimens with normal results in greater than 10 percent of specimens
  - Monthly and quarterly pathology specimens requiring repeat examination(s), listed by submitting physician
  - Quarterly and annual preventive medicine deficiencies
  - Mammography, by patient and age
  - Well woman examinations not performed
  - Immunizations overdue—childhood and adult boosters
  - Prenatal visits missed
  - Monthly anaphylaxis ED presentations by physician
  - Monthly unscheduled return to OR by physician, with reason stated
  - Monthly readmissions to hospital within thirty days of discharge from inpatient status
  - Monthly post-operative complications by physician, with reason stated
- *Quarterly listings of patients identified with the following diagnoses:*
  - Alzheimer's
  - Asthma
  - Atrial fibrillation
  - CHF
  - Chronic low-back pain
  - CVA
  - Depression
  - Diabetes
  - Fibromyalgia
  - GERD
  - Hypercholesterolemia
  - Hypertension
  - Inflammatory bowel disease
  - Irritable bowel disease
  - Lupus
  - Migraine
  - Myocardial infarction
  - Neoplasm
  - Obesity
  - Osteoporosis
  - Pregnancy
  - Renal failure
  - Rheumatoid arthritis
  - Stroke
  - TIA
  - Tonsillitis

The above report will enable the medical director and medical management/case management team to more closely coordinate appropriate resources for the patients and to reduce uncoordinated care for these patients as they often have concomitant disorders.

ADDITIONAL MONITORING REPORTS

- Monthly—Hospital admissions by DRG, by patient, by age, by admitting physician.
- Monthly—Length of stay by DRG, by patient, by age, by admitting physician.
- Monthly—Percent diagnosis seen by gender, by age, by zip code.
- Quarterly—Specialist peer comparisons, by billed charges, by specialty, by diagnosis.
- Quarterly—Admits for surgery by physician, by length of stay (LOS), by day of the week, by principal procedure:
    - Knees
    - Hips
    - CABG
    - Shoulders
    - Spine surgery

The above reports are helpful in determining who sees the patient more frequently with the same outcome for the same diagnosis with a higher or lower cost basis per diagnosis as related to visit frequency.

- Visits per-thousand-members-per-year (PTMPY)
- For the following diagnoses, encounters should each be tracked across the IPA or PHO, collectively, grouped into two reports: by PCPs and by Specialists:
    - Alzheimer's
    - Asthma
    - Atrial fibrillation
    - CHF
    - Chronic low back pain
    - CVA
    - Depression
    - Diabetes
    - Fibromyalgia
    - GERD
    - Hypercholesterolemia
    - Hypertension
    - Inflammatory bowel disease
    - Irritable bowel disease
    - Lupus
    - Migraine
    - Myocardial infarction
    - Neoplasm

- Obesity
- Osteoporosis
- Pregnancy
- Renal failure
- Rheumatoid arthritis
- Stroke
- TIA
- Tonsillitis

In conclusion, it is necessary to have the claims data flowing through a centralized repository to monitor all these items on a routine basis. It is also helpful to have a full-time nurse manager in the department who will run these reports and note any aberrances. This nurse should have a Certified Case Manager (CCM) background, loads of experience, and the actual CCM credential. Once the reports are prepared and reviewed by the nurse manager, he or she can prepare synopsis reports for the appropriate committees to review at the committee level and at the board level. It would be wise to copy your marketing committee with these periodic reports as well for their review and strategic planning purposes.

10 Provider Organization Credentialing and Privileging

Managed-care organizations consist of three primary areas: Health Maintenance Organization (HMO), Physician Hospital Organization (PHO), and Preferred Provider Organization (PPO). Additional categories of contracted services within the managed-care environment include Independent Practice Association (IPA), Management Services Organization (MSO), and Accountable Care Organization (ACO).

These entities all assume liability for the accuracy and thoroughness of the credentialing verification process and how this information is used in negotiating managed-care contracts with payers, employers, third-party audits, and the public. It is extremely important that the credentialing verifications conducted by Medical Staff Services Professionals (MSPs) within the managed-care arena are done accurately to reduce the risk of liability exposure. Designing, implementing, and maintaining strong credentialing practices also ensures that participating members in specific health plans are provided care by only the most qualified and competent practitioners.

The example text below is representative of what the contracting covenants usually say that gives rise to the liability for doing this correctly, and the assumptions that give rise to both vicarious and ostensible agency[1] theories of liability.

TYPICAL MANAGED-CARE PROVIDER ORGANIZATION REPRESENTATIONS AND WARRANTIES

a. Provider Organization represents and warrants that only properly credentialed and privileged participating physicians and allied health providers (hereinafter, "Represented Providers") will be allowed to provide covered services.
b. Provider Organization and each Represented Provider represent and warrant that the information set forth in the Medical Staff Credentialing and Privileging Data Set or other similar report furnished to payer is true and correct.
c. Provider Organization shall promptly notify Network of any changes in the information contained in the Medical Staff Credentialing and Privileging Data Set within thirty (30) days of such change.
d. Provider Organization represents and warrants that it is authorized to negotiate and execute contracts on behalf of its Represented Providers and will provide evidence of authority upon request.

e. Provider Organization shall use its best efforts to encourage compliance of Represented Physicians' agreement to abide by the terms of this Agreement and will provide evidence of it upon written request.
f. Health Plan makes no representations or guarantees concerning the number of Participants it can or will refer to Provider Organization under this Agreement.*
g. Health Plan disclaims any liability for credentialing and privileging decisions made by the Provider Organization and makes no representations or guarantees about credentialing decisions under this Agreement.
h. Health Plan will make best effort to market Represented Providers to client payers. Provider Organization shall have first right of review and right of refusal on any and all marketing documents that identify the Provider Organization specifically.

Courtesy: Maria K. Todd, Mercury Healthcare Advisory Group, Inc.

In 1991, the National Committee for Quality Assurance (NCQA) released its very first credentialing standards for HMOs. This action prompted many organizations throughout the country to examine their existing credentialing processes to determine what changes were needed to meet these stringent standards. The NCQA standards cover the entire scope of the operations of a Managed Care Organization (MCO) (e.g., policies, procedures, primary sources for verification elements, provider network directory, accepted sources to verify credentials, timelines for processing applications, etc.).

For the most current and up-to-date credentialing and privileging and widely accepted (and expected) best practices, readers are directed to the NCQA website, www.ncqa.org, for authoritative information.

The NCQA accreditation process is an interactive survey system (ISS) and is an automated desktop review system that allows the organization seeking accreditation to submit documentation of intended compliance over a secure electronic line. NCQA staff work with the organization's MSP to determine compliance for each standard. This automated process provides for compliance adherence to be completed prior to the site review, with the exception of the hands-on file audit.

My intention in this chapter is to provide readers with an overview of credentialing requirements and the credentialing process, including delegation of some or all of the credentialing activities.

At Mercury Healthcare International, we applied these standards across our provider network in the United States and ninety-five countries where healthcare professionals are presently privileged in our network. I have also used these checklists and standards to develop more than 150 provider organizations, including IPAs, PHOs,

* In a medical professional liability context, the ostensible agency liability doctrine is often used to hold hospitals liable for the acts of independent contractor physicians who work in emergency and operating rooms. There are two reasons for this approach. First, a provider organization environment creates a likelihood that patients will look to the organization rather than the individual physicians for care, primarily because of network steerage implications. Second, in many situations, a provider organization presents a physician as its member of some sort, independent or employed.

MSOs, Clinics without Walls, and now our trademarked term of art, the globally integrated health delivery system®.

Readers are also directed to consult the trade association for credentialing and medical staff services, the National Association Medical Staff Services (NAMSS; www.namss.org), to learn more about the technical aspects and principles of designing and implementing compliant credentialing and privileging systems, as well as obtaining knowledge about supplemental documents (policies, procedures, guidelines, etc.). For more personal and direct assistance, you may contact our office at Mercury Healthcare (www.mercuryhealthcare.us) as we currently include two NAMSS-certified and nationally recognized experts in the field of credentialing, privileging, and medical staff services who hold appointed positions in the Mercury Healthcare Advisory Group.

For those individuals or organizations interested in learning more, a quick search on the Internet will identify many organizations and software vendors that can assist the provider organization with electronically supporting the management of the credentialing system by supplying data, forms, software, or services capable of tackling all or many of the steps involved in the credentialing function.

PROVIDER EXPECTATIONS

Each provider organization is responsible for establishing criteria for participation within the organization, based on the needs of the members, clients, and the standards of the contracting entities. If you build it any other way, you have not met the first requirement of the market, which is to build a product that solves a problem for the customer. You just build the baseball diamond in the field of dreams … and the customers will not be arriving in a bus anytime soon.

PRACTITIONER REQUIREMENTS

There definitely will be some variation on the specific criteria required, but the basic credentialing elements for practitioners are likely to include the following:

- Valid and current licensure
- Clinical privileges at a hospital
- Valid and current Drug Enforcement Agency (DEA) Certificate
- Valid and current Controlled Dangerous Substance (CDS) Certificate, if applicable
- Appropriate and highest level of education and training: that is, graduation from an approved program recognized by the Accreditation Council for Graduate Medical Education (ACGME) and completion of an appropriate residency or specialty program
- Board certification (if specified by the practitioner or required by the organization or contracting agency)
- Appropriate documentation of work history and gaps
- Valid and current medical liability insurance
- History of medical liability claims and settlements

- Sanctions, restrictions, or limitations in scope of practice, as defined by the State Board of Medical Examiners or licensing agent
- Medicare and Medicaid sanctions
- Application with attestation

Regardless of the standards set (e.g., board certified, two years of experience, $2 million of medical liability insurance, etc.), the organization must put a system in place that ensures that its practitioners meet these standards before they are accepted as an active provider within the network. This system or process is commonly referred to as credentialing.

This term *credentialing* has been a bit confusing for providers in foreign countries establishing provider organizations for the purpose of medical tourism or international health access, with the intention of contracting with insurers, employers, third-party administrators, or other payer organizations such as governments and charitable missions, because many of the data elements are foreign to them. Also the verb *credentialing* has been confusing because they "already have credentials."

In these settings, I have had to change my idiomatic reference to *vetting* and *credentials verification process* or *primary source verification*, which are more easily understood in other countries outside the United States.

NON-PHYSICIAN AND PROVIDER REQUIREMENTS

There is no universal definition for the types of healthcare practitioners who are referred to as allied health professionals (AHPs) (e.g., dentists, chiropractors, psychologists, podiatrists, etc.). In 1997, the NCQA required some non-physician practitioners to be credentialed. MCOs are encouraged to review state statutes and other applicable standards to determine who and what is required to maintain compliance regarding the management of the non-physician practitioner. The provider organization is not required to credential practitioners who practice exclusively within another organization, such as the in-patient hospital setting, or free-standing facilities such as mammography centers, urgent care centers, or surgicenters because they are under contract with that organization and have no independent relationship with the provider organization.

The policies of the MCO must define the scope of practitioners to be covered, must establish criteria and the primary source verification that will be used to meet these criteria, and must delineate the process that will be used to make informed decisions and whether there exists any delegated credentialing or recredentialing arrangements.

The credentialing verification process for non-physician practitioners is much like that of physicians. There are differences in the requirements and, therefore, in the verification of select data. For example, chiropractors are not board certified and do not require DEA or CDS certificates. CDS certificates are also not applicable for dentists, but DEA certificates may be applicable, depending on the state in question. For the specific differences among the professional groups and requirements of each, the reader is directed to the NCQA Surveyor Guidelines (www.ncqa.org), which outline what is required for each type of practitioner and where to access such information. These guidelines are available for purchase and there may be different documents you will need, depending on the provider organization you may develop.

The NCQA is clear about the requirements for credentialing organizational providers. The provider organization is to have

1. Policies and procedures that speak to the initial and ongoing assessment of organizational providers with whom it intends to contract, including hospitals, home health agencies, skilled nursing facilities, nursing homes, and free-standing surgical centers.
2. Confirmation by primary source verification that the provider is in good standing with state and federal regulatory bodies, and has been reviewed and approved by an accrediting body. If the provider has not been approved by an accrediting body, the provider organization must develop and implement standards of participation.
3. Developed and implemented a policy and procedure and standards to "recredential" the provider at least every three (3) years by confirming that he or she continues to be in good standing with state and federal regulatory bodies and the appropriate accrediting body.

CREDENTIALING PROCESS

The credentialing process is just that: a very serious, step-by-step, tedious, and meticulous, detail-oriented process. A well-designed and effectively implemented credentialing system consists of a series of activities or steps that lead to a decision to accept or reject a practitioner's application to participate in an MCO as a health care provider.

NCQA Standards require each MCO to designate a Credentialing Committee that will make recommendations regarding the professional qualifications of its practitioners. These same standards also govern what the composition of the Credentialing Committee will be. The committee's goal is to ensure that the network's primary care and specialty services are represented appropriately.

Liability can be created for prejudice, refusal to deal, restraint of trade, discrimination, and a litany of other charges a disgruntled and rejected applicant can raise if your policies and procedures are not fair, consistently applied, and your documentation not thorough and accurate. There is an even greater need to be able to demonstrate compliance that your organization followed the defined process when a competitor is on the committee to evaluate competence that then ultimately decides to reject an applicant's request for participation.

A simple credentialing process is outlined below:

- *Application.* Practitioners expressing an interest in participation with the provider organization, and/or practitioners who meet the provider organization's organizational needs and administrative requirements, are invited to apply. Each applicant completes an application (credentialing) form, including a signed release granting the provider organization access to key information.

Each application serves as a request for participation and is accompanied by a copy of the applicant's current professional license, current DEA registration, if applicable, and the face sheet of the applicant's current professional liability insurance policy. The application for membership must include specific information as required by the NCQA and a statement by the applicant regarding

- Ability to perform essential functions of the position
- Illegal drug use (special policies will be needed for physicians in recovery)
- Loss of license or felony convictions
- Loss of limitation of privileges or disciplinary activity
- Correctness and completeness of the application

- *Initial Screening.* Before proceeding with the next step, the MSP reviews the application to determine that it is complete. If it is complete and meets the basic qualifications set out in a screening policy, it is forwarded to the Chief Medical Officer (CMO) or Medical Staffing Committee, who reviews it to determine if a preliminary interview is warranted, and if the full credentialing process is to be initiated—that is, verification of credentials through primary sources and new provider site visit.
- *New Provider Site Visit.* The applicant is notified that a new provider facility assessment and medical record keeping process audit must take place, which is conducted during the time of primary verification of credentials, and prior to the presentation of the applicant's file to a Credentialing Committee. The site visit includes an assessment of a number of criteria for which the provider organization has set out acceptable performance standards, including
    - Physical accessibility
    - Physical appearance
    - Adequacy of waiting and examining room space
    - Availability of appointments
    - Adequacy of treatment medical record keeping, which looks at how the practitioner documents his or her care, how he or she uses the documented information, how the file is organized, and how member confidentiality is maintained
    - Quality of care, which is determined by examining medical records selected at random and comparing the care provided against provider organization standards of care
- *Primary Source Verification.* The NCQA stipulates that seven (7) criteria must be verified from the primary source because they identify the legal authority to practice as well as the relevant training and experience. Provider organizations may choose to use an external agency to collect information from the primary sources. (The last time I checked on costs for this service, it was running about $175 per physician, plus any fees charged by verification sources.) If this is the case, the provider organization has delegated this function of the credentialing process and must assume oversight functions. These criteria that require primary source verification include
    - Valid license to practice

- Status of clinical privileges at the hospital designated by the practitioner as the primary admitting facility
- Valid DEA & valid CDS, if applicable
- Education and training of practitioners (highest level of education)
- Board certification if the practitioner states on the application that he/she is board certified
- Current adequate malpractice insurance according to the managed care organization's policy
- History of professional liability claims that resulted in settlements or judgments paid by or on behalf of the practitioner

- *File Preparation.* Immediately following the initial screening, the file is prepared for presentation to a Credentialing Committee. A file is initiated for each applicant that includes
  - Completed credentialing forms
  - Results from the new provider site audit (facility assessment and medical record keeping process)
  - Primary source verification of key elements
  - Work history, including investigation and documentation of gaps
  - Information from the National Practitioner Data Bank (NPDB), the relevant State Board of Examiners, and sanction activity by Medicare and Medicaid
  - Any other data relevant to the credentialing of the applicant
- *Data Entry.* After all required data elements have been received, the individual practitioner's credentialing file is entered into the provider organization's credentialing database and readied for presentation to a Credentialing Committee. A database assists in tracking an application's stage in the process and further assists in ensuring appropriate review of time-sensitive material (e.g., license, DEA Certificate, etc.). NCQA Standards require that information being presented to the Credentialing Committee be no more than 180 days old at the time of committee review. Nothing in the file can be more than 180 days old at the time of the committee's decision.
- *The Decision.* The decision to accept or reject an individual's application is made by a Credentialing Committee. The committee should be comprised of a range of participating practitioners who are capable of bringing technical knowledge and current medical practices to the attention of the provider organization. At a minimum, the committee reviews the files of all applicants who do not meet the provider organization's basic qualifications, and they should review the list and select files of applicants who do meet basic qualifications. The confidential minutes reflect the decision of the committee and any relevant discussion pertaining to the decisions. In the United States, these confidential minutes are protected under many State and Federal Peer Review Statutes. All applicants are notified of the committee's decision. A description of the appeals process should accompany all decisions not to credential practitioners.
- *Recredentialing.* MCOs are required to have an ongoing and current recredentialing process that is completed at least every three (3) years. In

considering whether to renew the practitioner's status with the health plan, the provider organization reviews information from the following sources:

- NPDB, State Board of Medical Examiners, and Medicare and Medicaid programs regarding sanction activity or practice limitations
- Member complaints and satisfaction results
- Quality improvement and utilization management activity reports, including review of case outcomes and patient satisfaction metrics
- Medical record reviews and facility site visit results: all primary care physicians, obstetrician/gynecologists, and high-volume specialists should have site visits every two (2) years, although it is more preferable to perform this every year

In addition to the above information sources, the provider organization must secure an attestation from the practitioner regarding his or her ability to perform the essential functions of the position, and assert one way or the other if he or she uses illegal drugs (I have never seen a positive response indicated on the attestation regarding this issue in my entire career!).

DELEGATED CREDENTIALING

Provider organizations can and do delegate all or some aspects of the credentialing process to outside organizations. While it is perfectly acceptable to delegate credentialing activities, the provider organization *does not relinquish responsibility for this function* and must continue to provide oversight. The first step in oversight is to evaluate the delegated entity's ability to perform the activities being delegated to your organization standards. The next step is to ascertain how "old" their information is.

Why I bring this up is because I have had the displeasure of using a Colorado credentials verification organization that was, at the time, accredited. They charged me $75 per physician, plus fees, but essentially resold us information in their files. So, we were paying for fees that had already been paid, not pass-through expenses. (Note to self: Get receipts for the work and expenses they do "for your organization, specifically.") Many of the data files they provided us were nearly two years old and, in some cases, older. They simply churned the database. What a racket! The entire job had to be redone; but when we argued that we did not want to pay for the garbage they gave us, they pointed to the language in their contract that was intentionally ambiguous about the currency of data. Do not let this mistake happen on your watch!

Upon determining that the delegated entity can successfully carry out the delegated function, the next step is to create a mutually agreed-upon document that describes the provider organization's responsibilities vis-à-vis the delegated entity.

The document should articulate the delegated activities, the process by which the provider organization will evaluate the delegated entity's performance, and how the provider organization will proceed if the delegated entity does not fulfill its obligations. Even with delegation, the provider organization retains the right to approve or reject individual practitioners based on quality-of-care issues. Finally, the provider organization must annually evaluate whether the delegated agency is conducting its activities according to the preestablished standards, to determine whether the contract will be renewed.

One of the more commonly delegated activities in credentialing is the primary source verification of select practitioner qualifications. Meeting the demand for this service are organizations known as Credentials Verification Organizations (CVOs), which will verify a practitioner's credentials for a set price. Since 1996, the NCQA has been reviewing and certifying CVOs for their compliance with their standards for the Certification of CVOs, and with its Credentialing Standards for Provider Organizations. When a CVO is certified by the NCQA, it means that the provider organization is exempt from the due-diligence oversight requirements specified by the NCQA for all the verification services for which the organization has been certified.

Certification does not guarantee satisfaction with the product or service, but it does offer the provider organization protection against scarce resources being spent on activities that cannot meet the rigorous demands of NCQA accreditation. The provider organization must be cautious in contracting with a CVO—it is possible for a CVO to be certified for some but not all of the ten essential elements specific by the NCQA. For those elements where certification has not been achieved, the provider organization is responsible for providing oversight.

I hope that this high-level summary of credentialing processes for the provider organization has conveyed the importance of the proper management of this important function. Please feel free to contact our experts with questions.

ACKNOWLEDGMENT

My deepest appreciation and thanks go to Ms. Donna Goestenkors, CPMSM , president of Team Med Global Consulting, a fellow Healthcare Consultant in the Mercury Healthcare Advisory Group and speaker, educator, author, mentor, and a past president of NAMSS, for her assistance with editing and revisions with this chapter.

11 Credentialing Committee's Assignment

What to Do and How to Do It

When an MCO (managed care organization) payor organization allows prospective providers into the network, it must perform diligent credentialing to attempt to ensure that the providers are expected to provide good quality care. This means that a process by which the participants are allowed to perform within the scope of their training is evaluated thoroughly by those granting the privilege to do so, with whom that clinician shall be affiliated. This process examines licensure, training, professional work history, qualifications, and even the malpractice claims history of the applicant.

Long-standing case law supports requirements that organizations that grant privileges to clinicians go beyond mere affirmations. One such case was the Nork case, which held the hospital liable for the actions of a staff physician because the hospital failed to verify the statements made on the physician's application, and simply relied on his truthfulness. Hospitals are not the only ones being challenged now for this, but so are IPAs (Independent Practice Associations), PHOs (Physician Hospital Organizations), and MSOs (Management Services Organizations) that grant privileges to treat patients of the network, as well as the HMOs (Health Maintenance Organizations) and PPOs (Preferred Provider Organizations) that contract with them. This opinion in the Nork case led to changes in The Joint Commission (TJC) standards that require that no clinical privileges be granted without prior verification of credentials. Other groups such as the Accreditation Association for Ambulatory Care call for review of training and peer review based on education, training and experience, and current competence of practitioners. Similarly, the National Committee for Quality Assurance (NCQA) has guidelines for credentialing within HMOs that a contracted IPA, PHO, or MSO would also have to follow.

A program of credentialing/recredentialing, whereby credentials should be reviewed every two years based on their original effective date with the IPA, should be developed. When it comes time to recredential (usually one or two years after acceptance into the IPA), recredentialing information should be requested by certified mail, return receipt requested. Note that even where rigorous credentialing exists, there are no guarantees of protection from liability arising from the negligent acts of a participating clinician.

In developing a credentialing policy, it would be helpful to incorporate a two-part policy containing a mechanism to review and categorize privileges to perform specific procedures and also to investigate whether a specific provider has demonstrated the necessary clinical competence to perform the procedures.

Some of the problems that can undermine a credentialing policy and actually create liability are failure to verify gathered information, failure to delineate privileges, and failure to follow up on data gathered.

Many groups perform what I call "social credentialing." This is a task that involves one or two people reading the responses on an application of someone well known to them professionally. After all, they send patients there on referrals, eat lunch with them, golf with them, etc. What's to examine?

Many groups newly formed, resist the need to establish well-planned and implemented credentialing policies that could prove legitimately defensible in the selection and retention process because they fear it will be too confrontational, and because they do not want to burden the clinician requesting privileges. They prefer to engage instead in social credentialing, and whether they are aware an applicant has active privileges at the participating plan hospital.

This is similar to purchasing a pig-in-a-poke in some cases. Applicant-provided references are also problematic in that colleagues do not usually say much more than "I have worked with him for so many years and feel he would be a great asset to our group." Many attorneys feel that this is downright prima fade evidence of negligence on the part of the MCO (Managed Care Organization) and that such requests for recommendations should come from the entity and be returned directly to the entity.

Another problem in credentialing often encountered is failure to follow up on data gathered. Should the entity granting privileges find something questionable in its data gathering, the courts usually consider this a warning notice to the entity. The failure to act on the information could be construed as entity negligence. This is usually referred to as nonfeasance, a failure to act when there is a duty to act.

DUE DILIGENCE IN CREDENTIALING

A good selection and retention program within any MCO should follow certain procedural guidelines, as there are numerous laws that impose responsibility upon those groups or entities that allow medical providers to "lay hands on" patients when the credentialing activities fall short of the necessary due diligence. These are usually addressed under tort or negligence law.

Some of the terms used in this area of law include *vicarious liability, respondeat superior,* and *ostensible agency.* All these theories of law are predicated on the fact that there is a network established and the fact that there should have been control exercised by one party over another party or the proximate cause of the selection of the individual to the injury. In other words, "If you had not chosen him to be in the network, he would not have been there to hurt me."

Traditional negligence law requires that the individual who harms a patient is responsible for his/her own actions when the harm is caused by breach of duty of care owed by the clinician to the victim. Why should an MCO that establishes the panel of providers be responsible for that clinician's actions? Three basic theories of law play a big role in the answer to the above: vicarious liability, ostensible agency theory, and master and servant or *respondeat superior* liability.

Vicarious Liability

Under the concept of vicarious liability, the MCO is liable for a provider's actions because of the special relationship that exists between them, even if the one on whom the vicarious liability is imposed is essentially blameless for the immediate harm. An example of this would be when a managed-care participating and credentialed physician requests time off and has another physician, who is nonparticipating and therefore not credentialed by the MCO, to provide backup coverage for him or her. In the event that the backup physician injures the patient, the vacationing physician may also be held liable because he or she allowed this physician to provide backup services when the backup may not have been qualified to do so. To this end, the MSO may also be held liable because they permitted the backup coverage by a nonparticipating, noncredentialed physician to see the patient, and therefore, the MCO can be found liable for the actions of the participating physician who selected and retained the backup.

A lawyer might argue that the MCO should have exercised more control over the participating physician by requiring that only other participating physicians may provide backup coverage to the managed-care patients. Therefore, both the vacationing physician and the MCO would be held vicariously liable for the backup's negligence. To dig a little deeper, the HMO or PPO that allowed the IPA to be the contracted providers, of which the vacationing physician is a member, might also be held responsible because of their lack of control over the IPA, who in turn should have exercised control over the vacationing physician, who should have been required to select a participating physician. This may be one of the reasons why we see some of the "hold harmless" provisions in managed-care contracts. Tort liability lawyers are simply doing their jobs when they dig for the deepest pockets in situations like these. A good credentialing procedure and related policies can help prevent these situations as long as they are not paper policies only. They must be enforced equally and fairly on everybody on a regular basis in order to be helpful.

Master and Servant Liability

Master and servant liability is commonly referred to by its Latin term *respondeat superior*, which is used to describe the responsibility of the master (employer) for the servant (employee). In managed-care relationships, most often you will see contractual language that attempts to reflect that no master–servant relationship exists, no employer–employee relationship exists. I am aware of several cases throughout the country where this clause in the contract has been tested. The argument routinely used is that the physician or other participant is following policies, procedures, and standards of care set forth by the organization, must be available for work 24 hours per day, 365 days per year, and that monies are paid by a fee schedule and therefore this is not a clear independent contractor relationship. Often, MCOs of all types will include independent contractor language in all contracts—whether IPA, PHO, MSO,

HMO, or PPO—so that the payor does not have to deduct FICA*, FUTA†, and withholding taxes, or pay workers' compensation when tendering payment for services rendered. The courts have, in the past, tested the merits of the case by the facts and not just the written words. In recent cases, they have found the MCO liable as master for the actions of a member. I am not suggesting that you change the participation agreements, but understand that you are not protected just because the agreement says you are supposed to be. For more information, review *Darling v. Charleston Community Memorial Hospital* (33 Iil. 2d 326, 211 N.E. 2d 253 (1965), cert denied, 383 US 946 (1966)).

Ostensible Agency

When the MCO is liable for the actions of its participating providers (and nonparticipating providers who have been allowed to see patients for whatever reason), the law imposes liability on the MCO for the actions of its members who are apparently acting for the MCO, and therefore are presumed to be under the MCO's control.

To explain this further, we need to understand the difference between an HMO or other health plan and an indemnity insurance plan. In the indemnity plan, you pay your premium and receive a contract or certificate that declares what is covered and what is not covered. There is no limitation as to where you may seek care, because the care is unregulated by the indemnity carrier. As long as a physician states that something is medically necessary and it falls within policy limitations and guidelines, the bill is paid at the contract rate. There is no "management of care," no provider panel, and no incentive to use a particular provider network.

By contrast, in an HMO or PPO setting, there is a panel because the mechanism is not that of insurance, but that of a health plan. In a health plan, the insurance mechanism of dealing with financial risk for the cost of claims is married to a delivery mechanism, and the benefactors of the plan are required to use the specific network in order to maximize benefits through the plan.

Therefore, it would stand to reason that those benefactors would be entitled to some protection in the event of being injured by the negligent actions performed by any member of the designated panel of providers that they are required to use.

Some of the traditional credentialing requirements should include the following:

- Physician licensure
- Board certification
- Professional liability coverage
- Professional history and references
- Hospital affiliations

The IPA should also access information from the National Practitioners' Data Bank (if applicable) or other department of professional regulation queries, and take into consideration patient satisfaction surveys, quality assurance concerns, cost

* Federal Insurance Contributions Act.
† The Federal Unemployment Tax Act.

utilization profiles, PCP change profiles (if applicable), and the individual provider's commitment to the managed-care team philosophy. In addition, the credentialing committee should take into consideration the provider's open/closed status of the ability to add new patients, chart review scores, ER* utilization (if applicable), sanction history, coding practices, cost rankings, and C-section rates (if applicable).

The credentialing procedure can raise problems of its own with regard to conspiracy theories, defamation, civil rights violations, and possible antitrust actions regarding restraint of trade. You should know that the Health Care Quality Improvement Act of 1986 (P.L.99-160) has provisions of law that provide some protections, especially with regard to antitrust concerns. These protections are granted as long as when the MCO prefers an adverse decision against an applicant, the decisions can be demonstrated to be both (1) based on firm belief that the applicant would have posed a quality problem and (2) that they offered the applicant an opportunity for due process. Another stipulation is that the conditions that caused the applicant to be denied are, in certain instances, reported to the National Practitioner Data Bank.

Any MCO that enters into affiliated relationships with physicians, where the physician has applied for clinical privileges or appointment, is granted access to the databank. Data available include licensure actions, malpractice claims and settlements, and privilege and membership restrictions by other reporting agencies. In addition, reportable decisions are described as those that adversely affect the privileges of a practitioner for longer than thirty days, including reducing, restricting, suspending, revoking, or initially denying privileges. Failure to report vitiates the protections provided by the Act.

STRUCTURING A GOOD CREDENTIALING POLICY

1. Begin with a clear and concise statement of policy existence of a process whereby credentials are reviewed and verified through primary source verification; this is mandatory for affiliation with the MCO. Also state the recredentialing policy and the time element involved.
2. Next, develop the written policy that defines what will be required for information submitted for review. Also, make sure to include some policy on due process for those denied privileges, right to attorney representation, if desired, and how denials will be communicated to the applicant. In addition, spell out how confidentiality will be maintained.
3. Develop specific form letters to be used, a standard application, and a reference request form. Attach and incorporate them into the organization's credentialing policy as exhibits.
4. Finally, stick to your policies and procedures. No matter how time consuming, it is no defense to say that the organization was too busy to carry out the necessary due diligence in matters of privileging and retention in an MCO of any type.

* Emergency Room.

In the next few pages, I have listed some of the elements most commonly found in credentialing applications. You may want to use this listing to create a multi-page application for privileges and membership for your MCO. Also, check with your local hospital credentialing expert, as your state may have enacted a law that requires a specific form to be used for all credentialing as a standard application. You will need to verify if the application is required for use by all MCOs, or HMOs and PPOs only, or what the rules are.

If you are creating your own application, the following layout will help.

Remember to state in boldface type at the top of the document your design:

> "This application is for credentialing purposes only. It is not intended to guarantee contracted provider status or membership in the GreatCare IPA (or PHO or MSO)."

You may wish to title your document

PROVIDER APPLICATION FOR CREDENTIALING

[I always like to include the following instructions:]

1. Please provide *all* requested information.
2. Please type or print.
3. Use "N/A" throughout this application for "not applicable."
4. Attach a copy of your Curriculum Vitae and any other required materials requested.
5. Sign any releases for information.
6. Sign this application and return all materials in the enclosed envelope.

General Information

1. Full name
2. Federal tax identification number(s)
3. Federal narcotics registration number(s) (if applicable)
4. Social security number
5. Medicare provider identification number, UM number
6. Current office, principle address, phone, fax, answering service
7. Alternative office, address, phone, fax, answering service
8. Home address and telephone number, fax
9. Date of birth, birthplace
10. Medical or professional school, attendance dates from/to, degree conferred
11. If applicable, ECFMG® number, FMG* Clinical Clerkship location, dates from/to
12. Internship, location, type, dates from/to
13. Internship, location, type, dates from/to (if more than one)
14. Residencies, location, address, from/to, specialty

* Foreign Medical Graduates.

15. Fellowships, location, address, from/to, specialty
16. Teaching appointments, location, address, type, from/to
17. Military service, branch, from/to status or separation type

[The next questions should have room for free text answers.]

18. Under what specialties or subspecialties do you desire to be listed in the published provider list?
19. What is the complete name of your practice/group?
20. Have you ever been convicted, whether as a result of a guilty plea, a plea of nolo contendere, or verdict of guilty, of a felony, any offense involving moral turpitude, or any offense related to the practice of, or the ability to practice, medicine or the related healing arts? ☐ Yes ☐ No
21. Do you have any physical or mental condition, including alcohol or drug dependency, that may affect your ability to fulfill your professional responsibilities? ☐ Yes ☐ No
22. Have you ever had such a condition in the past that is now resolved without need for continuing therapy or medication? ☐ Yes ☐ No
23. Have you ever been hospitalized or received any other type of institutional care for such a condition during the past ten (10) years? ☐ Yes ☐ No
24. Are you currently taking medication or under other therapy for a condition that could affect your ability to fulfill your professional responsibilities if the medication or therapy were discontinued today? ☐ Yes ☐ No
25. Have you ever had privileges with any managed-care organization reduced or terminated for any reason? ☐ Yes ☐ No

[The above questions 20 through 25 should be asked with a place for a "Yes" or "No" response. Furthermore, if there is a "Yes" response, details should be requested (required).]

Licensing/Certifications/Registration Numbers

26. List the name of all states in which you are licensed, registered, certified, or otherwise authorized to practice your specialty.
27. List board certifications (if applicable) by specialty, and include certification date(s) and recertification date(s).

Hospital Privileges

28. List the name/city/state of all hospital(s) at which you currently hold active, courtesy, or provisional staff privileges, and list the privilege level and each department in which you have clinical privileges. Attach additional pages if necessary.
29. List the name/city/state of all hospital(s) at which you previously held active, courtesy, or provisional staff privileges in the past ten (10) years.

Professional Liability Coverage (at the time this application is completed)

30. Name of your professional liability insurance carrier, policy number, policy period.
31. Have you ever been accused of malpractice, or has a suit for any alleged malpractice ever been brought against you? If yes, please provide complete details for each occurrence, including allegations, dates, amounts of suits, and amount of any final settlements or judgments on an attached sheet.
32. Has your professional liability insurance ever been canceled, premiums surcharged, or its renewal refused?

Disciplinary Actions

33. Have you ever been subject to disciplinary review or action, or are pending action, by any of the following:

    State Medical/Professional Licensing Board ☐ Yes ☐ No [State]
    County or State Medical/Professional Society ☐ Yes ☐ No [State]
    Hospital or Medical Staff ☐ Yes ☐ No [Hospital]
    Drug Enforcement Agency ☐ Yes ☐ No [State]
    Military ☐ Yes ☐ No [Site]
    Centers for Medicare and Medicaid Services (CMS) ☐ Yes ☐ No [State]

[If "Yes" answers are given, make sure to obtain details and follow up on them.]

Professional References

34. Please list three professional references with addresses from within the discipline in which you are licensed, registered, certified, or otherwise authorized to practice.
    - If you have practiced in your current community for less than three (3) years, please make sure that one of the references is from your previous community.
    - Also please note that by signing the release enclosed with this application, you are providing (insert MCO name) with your authorization to verify professional references that may have been provided to others (i.e., hospitals) when being credentialed there.
    - If we cannot obtain these references, it may cause your application to be delayed, pended, or even denied.
35. We would appreciate you attaching a current black-and-white photograph to be used in any future physician brochures.

Office Information

[Here, you may want applicants to provide the following office and practice information, which will be helpful in the orientation of their office staff.]

36. Name of office manager or contact person.
37. Billing address.
38. Days/hours in office.
39. What are your provisions for after-hours care?
40. Do you have 7 days-a-week, 24-hour backup or call coverage?
41. Indicate who you use for backup coverage.
42. Indicate the age range of patients you will treat.
43. If you are a family practitioner or general practitioner, do you handle pediatrics? Obstetrics? Gynecology? Geriatrics?
44. Do you use non-physician practitioners in your office practice? ☐ Yes ☐ No
45. If yes, are they covered by your professional liability insurance? ☐ Yes ☐ No
46. What other licensed personnel work in your office?
47. Please indicate to whom you refer most of your patients for the following:
    - Surgery ________
    - Obstetrics ________
    - Cardiology ________
    - Urology ________
    - Ophthalmology ________
    - Orthopedics ________
    - Gastroenterology ________
    - ENT ________
    - Oncology ________
    - Psychiatry ________
    - Family/general practice ________
    - Lab and pathology ________
    - Radiology ________

[Note: Only ask about the specialties that are relevant!]

Finally, add some language for the releases and signatures such as

> "I hereby certify the above information is true and accurate, and I understand that any material misstatements in or omissions from this application shall constitute good cause for denial of eligibility, or result in a later termination as a contracted or employed provider of care for (insert MCO name)."

Also, it may help to use the following as a release of information language for reference checks:

> "I have applied to [insert MCO name] for eligibility to contract as a member of the [insert MCO name], and hereby authorize [insert MCO name] and its representatives to consult with the administration and members of the medical staffs of hospitals, institutions, professional licensing/registration/ certification bodies, professional liability insurance carriers, and professional organizations with which I have been associated and with others who may have information bearing on my professional competence, character, and ethical qualifications. I hereby further consent to the inspection by

[insert MCO name] and its representatives of all documents and information that may be material to an evaluation of my professional qualifications and competence.
By signing this authorization, I release from liability all representatives of [reference name], as well as all representatives of hospitals, institutions, professional licensing/ registration/ certification bodies, and professional organizations for their acts performed in good faith and without malice in connection with both the exchange of information as consented to above, as well as in connection with evaluating my application, my credentials, and my qualifications. A photocopy of this authorization is to be accepted with the same authority as this original."

Still another release may be necessary to obtain information about professional liability insurance and a Certificate of Coverage. This one is simple and needs to contain some statement such as

"I hereby authorize [insert insurance company name] at [insert address] to issue a copy of my current Certificate of Coverage to [insert MCO name], including any notification of any changes in my policy and notification of any future actions that may be filed against me."

It is also a good idea to create a document checklist for each credentialing file. One that I use for my new MCO start-up clients looks like the following list:

- Two copies of the Provider Services Agreement (both completed and signed)
- Credentialing application
- Information sheets
- Medical license
- DEA Certificate
- Malpractice insurance declaration pages showing amounts of coverage
- Board certification certificates
- Peer recommendation (letters)
- Residency certificates, if not board certified
- Internship certificates, if not board certified
- Medical school graduation certificate
- Photograph

With this much information and the true primary source verification, you should be well on your way to developing a good credentialing policy and procedure for any organization, whether an IPA, PHO, or MSO.

12 Antitrust Compliance Task Force

*Understanding Antitrust Concerns for Provider Networks**

Note: This chapter is designed to provide general information about the antitrust laws, and the Department of Justice and Federal Trade Commission's guidelines on antitrust enforcement policy in healthcare. The memorandum does not constitute specific legal advice, nor does it represent an official position for Maria K. Todd, the Mercury Healthcare Companies, or CRC Productivity Press as the publisher. For specific questions about the contents in this memorandum, please contact your local healthcare attorney.

This chapter provides a general summary (with many verbatim excerpts) of these guidelines, which were provided in the form of policy statements issued jointly, and subsequently amended, by the United States Department of Justice and the Federal Trade Commission (FTC) between September 1993 and September 1996.

The IPA (Independent Practice Association), PHO (Physician Hospital Organization), and MSO (Management Services Organization) groups that have been forming in the past few years continue to request information about the federal guidelines for antitrust enforcement policy in healthcare. I find it uncanny that the hundreds who have attended both my contracting seminars and IDS Development seminars state that they are actively engaged in IDS participation or development and have never heard of these documents before. This frightens me for their sake as there is so much increased prosecutorial activity going on these days. You cannot take antitrust rules for granted or make assumptions about prosecutorial actions.

As a paralegal, I cannot provide specific legal advice as to whether your business activity is in compliance with federal antitrust law and these guidelines. Moreover, this chapter and the policy statements mentioned do not include any analysis of enforcement policy resulting from legal challenges. This antitrust case law is equally important and relevant when considering any type of business activity. Therefore, I urge you to contact your local healthcare attorney to provide you with further guidance whenever you are contemplating a business venture that

* For specific questions about the contents of this chapter, please contact a qualified health law or business attorney.

may fall under the purview of the Department of Justice and the Federal Trade Commission's jurisdiction.

WHAT ARE THE ANTITRUST LAWS?

First enacted in the 1890s, the purpose of the antitrust laws was to protect consumers against anti-competitive business activities. These activities, especially the creation of monopolies, enabled businesses to charge higher prices to consumers (price fixing). The law simply stated that "every contract combination in the form of trust or otherwise, or conspiracy, in restraint of trade or commerce" was illegal, and every person who shall "monopolize, or attempt to monopolize, or combine or conspire with any other person" was in violation of the law.

Although this law was specifically intended to be a consumer protection measure, ironically, at times, consumers and the marketplace are best served through combined business arrangements and the sharing of information, which may otherwise be interpreted as a monopoly or restriction on trade. This is the paradox of the antitrust laws. Sometimes, one very large business providing goods or services for everyone "gouges" consumers, and the federal government needs to step in (such as the break-up of AT&T or Standard Oil). Other times, companies merging together can produce business efficiencies and savings to consumers (such as the recent proposed mergers of some of the "Baby Bell" telephone companies).

The growth and dissolution of companies, as dictated by the antitrust laws, is cyclical, and the interpretation and application of the antitrust laws is constantly evolving. In its application, the government always takes into consideration the nature of the current marketplace, the effect of this marketplace on consumers, and the pricing of goods or services.

Consequently, because some current combined business activities in the healthcare industry have resulted in efficiencies in healthcare delivery and cost-savings to the patient, the United States Department of Justice and the Federal Trade Commission (DOJ/FTC) have provided some guidance as to what activities would not run afoul of the antitrust laws in this new healthcare delivery environment.

TERMS USED IN THE GUIDELINES

The DOJ/FTC have developed their own terms to describe conduct in the healthcare industry, and they use these terms to describe what conduct will be considered legal as opposed to illegal.

Key terms are as follows:

Physician network joint ventures: Physician-controlled ventures in which the network's physician participants collectively agree on prices or price-related terms and jointly market their services. They will be referred to as "physician networks." They include Independent Practice Associations (IPAs), Preferred Provider Organizations (PPOs), and other arrangements. At least some, if not all, of the physicians are assumed to be in independent practices as opposed to group practices where the physicians have merged their practices. These networks are not fully integrated; they are either partially integrated or unintegrated. Therefore, price agreements among physicians in the network amounts to per se illegal price fixing unless the network is sufficiently integrated to qualify for rule of reason analysis.

Multi-provider networks: Ventures among providers that jointly market their healthcare services to health plans and other purchasers. They include networks among competing providers as well as networks of providers offering complementary or unrelated services. A Physician Hospital Organization (PHO) is a multi-provider network. At least some, if not all, of the providers in the network are independent as opposed to being part of the same business entity. Normally, these networks are not fully integrated; they are partially integrated or unintegrated.

Antitrust safety zones: Categories of physician networks that will not be scrutinized by the DOJ/FTC. The safety zones are defined in the guidelines. However, being within a safety zone does not bar private parties from bringing suit. Safety zones are not laws, and it is possible that a court may disagree with the DOJ/FTC. There are no safety zones for multi-provider networks. Absent extraordinary circumstances, the agencies will not challenge activity that is identified as falling within an antitrust "safety zone," so long as all specific requirements of that "safety zone" are met.

For all of these identified antitrust "safety zones," providers can take advantage of the DOJ's expedited business review procedure, or the FTC's advisory opinion procedure as outlined in the Code of Federal Regulations. The agencies will respond to a business review or advisory opinion request within ninety (90) days after all necessary information is submitted by the providers who are considering joining business activities that may fall within a particular "safety zone." The agencies will provide an opinion as to whether the business activity under consideration falls within an antitrust "safety zone."

Per se violation of antitrust laws: Antitrust laws that consider naked agreements among competitors that fix prices or allocate markets as illegal on their face (that is, a per se violation of the antitrust laws).

Exclusive network: An exclusive physician network that does not allow its member physicians to belong to a competing healthcare delivery network, or which does not allow its members to contract with payors except on terms that the network accepts, or both. Exclusivity may be found to exist because it is expressly required, or it might be found to exist as an implicit agreement, meaning that the physicians do not deal with payors outside of the network or join other networks even though it is not an express requirement. An exclusive network is potentially threatening to competition because its physicians are not available to form competing networks or to participate in managed-care plans through arrangements made outside of the network; therefore, it is treated more restrictively under the antitrust laws.

Nonexclusive network: Physician network that allows its members to contract with competing healthcare delivery networks or to contract individually with payors on terms not accepted by the network. It is not as threatening to competition as an exclusive network, and therefore it is treated less restrictively. However, the DOJ/FTC is concerned that a nonexclusive network might actually be an exclusive network in disguise to take advantage of the less-restrictive antitrust treatment.

To determine whether a network is truly nonexclusive, the DOJ/FTC will look at a number of criteria, including (1) whether viable competing networks or health plans with adequate provider participation exist in the market, (2) whether members of the nonexclusive network actually contract with other networks or health plans or whether there is evidence that they are willing to do so, (3) whether providers in the nonexclusive network earn substantial revenue outside the network, (4) whether there is evidence that members of the nonexclusive network are "departicipating" from other networks, and (5) whether the members of the nonexclusive network coordinate their prices or terms of dealing with other networks or health plans.

Horizontal arrangements: Arrangements between competing providers, such as physicians in the same specialty. The antitrust laws are highly restrictive of horizontal arrangements, as it is believed that they are likely to be anti-competitive.

Vertical arrangements: Arrangements between providers that perform different services and do not compete, such as hospitals and physicians. The antitrust laws are less restrictive of vertical arrangements because they are usually meant to achieve efficiencies. It should be noted, however, that a vertical arrangement can also have horizontal dimensions. For example, when a hospital makes an arrangement with a physician network, the combined multi-provider network has both vertical relationships between the physicians and the hospital, and horizontal relationships among the physicians.

Physicians in a multi-provider network may not engage in price fixing simply because the multi-provider network has a vertical dimension. In addition, a hospital that employs physicians is in competition with independent physicians in the same specialty, and therefore arrangements between the hospital and those competing physicians are horizontal, not vertical.

Rule of reason analysis: The "rule of reason" analysis is undertaken by the Department of Justice in cases where competitors, for economic efficiency reasons, integrate in a joint venture. Such agreements, if reasonably necessary to accomplish the pro-competitive benefits of the integration, are analyzed under the "rule of reason" a doctrine developed by the Supreme Court of the United States (SCOTUS) in its interpretation of the Sherman Antitrust Act. The rule is that only combinations and contracts unreasonably restraining trade are subject to actions under the anti-trust laws, and that possession of monopoly power is not inherently illegal.

Under rule of reason analysis, physician networks would not be viewed as being per se illegal as long as two conditions are met:

1. The physicians' integration through the network is likely to produce significant efficiencies that benefit consumers; and
2. Any price agreements (or other agreements that would otherwise be per se illegal) by the network physicians are reasonably necessary to realize those efficiencies and fall within the antitrust "safety zone."

As with any other type of antitrust analysis, determining whether an arrangement is merely a vehicle to fix prices or engage in naked anti-competitive conduct, or a device to produce efficiencies and consumer benefits, is a factual inquiry that must be done on a case-by-case basis. Again, the agencies' analysis is to determine the arrangement's true nature and likely competitive effects.

FEDERAL GUIDELINES OF ANTITRUST ENFORCEMENT POLICY IN HEALTHCARE

On August 28, 1996, the United States Department of Justice (DOJ) and the FTC jointly released antitrust guidelines for the healthcare industry. These guidelines expand upon, and clarify, the guidelines that were first issued in September 1993 and later revised and expanded in September 1994.

During the past several years, the healthcare marketplace has been rapidly changing. During this time of tremendous change, there was a great deal of uncertainty concerning the federal government's antitrust enforcement policy. Prior to 1993, some believed that this enforcement policy actually was inhibiting mergers and joint activities that enhanced patient healthcare delivery and controlled the cost to the consumer. Consequently, in September 1993, the DOJ and the FTC released antitrust guidelines to provide some guidance to healthcare providers and hospitals so that they could enter into joint ventures, mergers, and other collaborative activities without violating the antitrust laws.

These original guidelines included six policy statements on hospital mergers, hospital joint ventures involving high-technology or other expensive medical equipment, hospital participation in exchanges of price and cost information, physicians' provision of information to purchasers of healthcare services, healthcare providers' joint purchasing arrangements, and physician network joint ventures.

At that time, the DOJ and the FTC made a commitment to issue expedited business review or advisory opinions in response to requests for antitrust guidance on specific proposed healthcare arrangements.

The 1994 guidelines, which superseded the 1993 statements, added new policy statements, including hospital joint ventures involving specialized clinical or other expensive healthcare services, providers' collective provisions of fee-related information to purchasers of healthcare services, and analytical principles relating to a broad range of healthcare provider networks ("multi-provider networks"). It also expanded the antitrust "safety zones" for several of the 1993 policy statements.

Since 1994, the DOJ and the FTC have gained much experience with arrangements involving joint provider activity and therefore have once again provided further clarification. The August 1996 guidelines, which are almost 150 pages in length, expanded the enforcement policy statement on physician network joint ventures and the more general statement on multi-provider networks.

These most recent guidelines from August 1996, which supersede the previous policy statements, are now comprised of nine policy statements, of which I review five of the antitrust enforcement policy guidelines that concern physician and other providers' activities, just so you can get a feel for how important it is to remain vigilant in your efforts to comply and document decisions according to competent legal counsel's guidance. Many of the provider organizations and consultants that have involved the wrath of the market and the DOJ and the FTC had done so because while their intent was to comply, and they filed the requisite paperwork, and even obtained private letter rulings at a cost of tens of thousands of dollars, they inadvertently (or strategically) changed course and direction right out of the safety zone they originally proposed or documented.

If a physician who was a shareholder in the 1990s was involved in one of these cases, he or she may also be enjoined from becoming a shareholder in your organization. This may mean that on credentialing and application paperwork, you might somehow also inquire *if* they are eligible to join your group.

Statement 4	Providers' Collective Provision of Non-Fee-Related Information to Purchasers of Healthcare Services
Statement 5	Providers' Collective Provision of Fee-Related Information to Purchasers of Healthcare Services
Statement 6	Provider Participation in Exchanges of Price and Cost Information
Statement 7	Joint Purchasing Arrangements Among Healthcare Providers
Statement 8	Physician Network Joint Ventures
Statement 9	Multi-Provider Networks

Let us examine these statements individually:

Statement 4: Providers' Collective Provision of Non-Fee-Related Information to Purchasers of Healthcare Services

The agencies have designated an antitrust "safety zone" where competing providers can collect and share non-fee-related information when they meet certain requirements. Providers can collectively share underlying medical data that may improve the mode, quality, or efficiency of treatment, and this activity is unlikely to raise any significant antitrust concerns. Statistics such as outcome data or practice parameters may be collected by providers, and they may collectively engage in discussions with purchasers about the scientific merit of that data.

This "safety zone" specifically excludes any attempt by providers to coerce a purchaser's decision making by implying or threatening a boycott of any plan that does not follow the providers' joint recommendation.

Statement 5: Providers' Collective Provision of Fee-Related Information to Purchasers of Healthcare Services

The agencies have designated an antitrust "safety zone" for the collective sharing of fee-related information by competing providers, when they meet certain requirements. With reasonable safeguards in place, competing healthcare providers can collectively provide to purchasers factual information concerning the fees charged currently and in the past, as well as other factual information concerning the amounts, levels, or methods of fees or reimbursement.

In order to qualify for this "safety zone," the collection of information to be provided to purchasers must meet all of the following requirements:

1. The collection is managed by an unaffiliated third party (such as a healthcare consultant or trade association).
2. Any information that is shared among or is available to the competing providers furnishing the data must be more than three (3) months old.
3. For any information that is available, there are at least five providers reporting data upon which each disseminated statistic is based; no individual provider's data may represent more than 25 percent on a weighted basis of that statistic; and any information disseminated must be sufficiently aggregated so recipients cannot identify the prices charged by any individual provider.

This antitrust "safety zone" does not apply to collective negotiations between nonintegrated providers and purchasers relating to any agreement among the providers on fees or other terms or aspects of reimbursement, or to any agreement among integrated providers to deal with purchasers only on agreed terms. Also, providers cannot collectively threaten or engage in a boycott or similar conduct to coerce any purchasers to accept collectively determined fees.

Also, specifically excluded from this "safety zone" are providers' collective provisions of information or views concerning prospective fee-related matters.

Statement 6: Provider Participation in Exchanges of Price and Cost Information

The agencies have designated an antitrust "safety zone" for provider participation in the exchange of price and cost information, when they meet certain requirements. Absent extraordinary circumstances, providers can participate in written surveys of prices for healthcare services, or wages, salaries, or benefits of healthcare personnel as long as

1. The survey is managed by an unaffiliated third-party (such as a healthcare consultant or trade association).
2. The information provided by the survey participants is based on data more than three (3) months old.
3. There are at least five providers reporting data upon which each disseminated statistic is based; no individual provider's data represents more than 25 percent on a weighted basis of that statistic; and any information disseminated is sufficiently aggregated so recipients cannot identify the prices charged or compensation paid by any particular provider.

   Any other exchanges of price and cost information that fall outside this antitrust "safety zone" will be evaluated to determine whether the information exchange may have an anti-competitive effect that outweighs any pro-competitive justification for the exchange. Such surveys, including public, nonprovider initiated surveys, may not raise competitive concerns. However, exchanges of future prices for provider services, or future compensation of employees, are very likely to be considered anticompetitive. It is illegal to exchange among competing providers price or cost information that results in an agreement among competitors regarding the prices for healthcare services or the wages to be paid to healthcare employees.

Statement 7: Joint Purchasing Arrangements among Healthcare Providers

The agencies have designated an antitrust "safety zone" for competing providers to enter into joint purchasing arrangements (such as for the purchase of computers or pharmaceutical products) when they meet certain requirements. These joint purchasing arrangements may allow the participants to achieve efficiencies that benefit consumers by lowering the cost of healthcare.

These joint purchasing arrangements are unlikely to raise antitrust concerns unless the arrangement accounts for such a large portion of the purchases of a product or service that the participants can effectively exercise market power in the purchase of the product or service; or the products or services being purchased jointly account for so large a proportion of the total cost of the services being sold by the participants that the joint purchasing arrangement may facilitate price fixing or otherwise reduce competition.

Absent extraordinary circumstances, the agencies will not challenge joint purchasing arrangements among healthcare providers, as long as two conditions are met:

1. The purchases account for less than 35 percent of the total sales of the purchased product or service in the relevant market.
2. The cost of the products and services purchased jointly accounts for less than 20 percent of the total revenues from all products or services sold by each competing participant in the joint purchasing arrangement.

Joint purchasing arrangements among healthcare providers that fall outside the antitrust "safety zone" do not necessarily raise antitrust concerns. There are several safeguards that joint purchasing arrangements can adopt to avoid some antitrust "pitfalls" and demonstrate that the joint purchasing arrangement is intended to achieve economic efficiencies, rather than to serve anti-competitive purposes:

1. Members should not be required to use the arrangement for all of their purchases of a particular product or service (however, voluntarily specified amounts are acceptable, so that a volume discount or other favorable contract can be negotiated).
2. Negotiations should be conducted on behalf of the joint purchasing arrangement by an independent employee or agent who is not also an employee of any participant.

3. Communications between the purchasing group and each individual participant should be kept confidential, and not discussed with or disseminated to other participants.

It is not necessary to open a joint purchasing arrangement to all competitors in the market. However, if some competitors excluded from the arrangements are unable to compete effectively without access to the arrangement and competition is subsequently stifled, this activity may raise antitrust concerns.

Statement 8: Physician Network Joint Ventures

The agencies have designated several "antitrust safety zones" for physician network joint ventures when they meet certain requirements. A physician network joint venture is a physician-controlled venture in which the network's physician participants collectively agree on prices or price-related items and jointly market their services (such as IPAs and PPOs).

USE OF THE MESSENGER MODEL TO NEGOTIATE AN AGREEMENT WITH A PAYOR

Note: These scenarios and determinations of legality are verbatim from Frequently Asked Questions provided by the DOJ. They are presented as informational and do not constitute legal or professional advice. Antitrust laws are always decided on case specifics.

Characteristics of the Arrangement

About 35 percent of the physicians in a community decide that they want to negotiate an agreement with a payor, but they want to avoid the bar on price fixing. They decide to appoint a third party to act as a "messenger" in negotiations with the payor. The messenger will collect information from each of the physicians about the fee range that the physician is willing to accept, and each physician will give the messenger authority to accept contract offers from payors that fall within or above the fee range desired by the physician. The messenger will share the fee range desired by any physician with the other physicians. The messenger will aggregate the information and create a schedule showing what percentages of physicians will accept contract offers at various fee levels.

The messenger will then present the schedule to payors and solicit offers. Offers that fall within or above the fee range of any physician will be accepted on the physician's behalf. Offers that do not fall within a physician's fee range will be forwarded to the physician, and the physician may accept or reject the offer. The messenger will not tell any physician whether other physicians intend to accept an offer or not, but will provide physicians with objective information about any offer, including how it compares with the offers of other payors and the meaning of contract terms. The physicians will not discuss any contract offers among themselves. Once the messenger obtains the responses of the physicians, that data is relayed to the payor, and contracts are finalized.

Legality of the Arrangement

Statement 9 defines this type of arrangement as the "messenger model" and says that it may be used, although it does not define a safety zone for it. These characteristics must be strictly adhered to in order to stay within the law. Statement 9 warns that variations may cause the arrangement to be price fixing,

The section on the messenger model in Statement 9 does not place any size limits on non-integrated groups that use it. The DOJ issued a pair of consent judgments involving Physician Hospital Organizations (PHOs), which say that the PHOs, both of which include very large percentages of physicians in their markets, may legally use the messenger model to arrive at fee arrangements with payors.[*] In addition, officials of the DOJ have publicly stated that they do not see a reason to place limits on the percentage of physicians that can participate in a messenger model arrangement.

The rationale for that position is that there is no agreement among the physicians, and therefore there is no joint venture subject to limits on its market power. However, if a messenger model arrangement involves more than an arrangement to arrive at fees, and also involves agreements among physicians to conduct their practices pursuant to certain terms, those agreements may make the arrangement a joint venture, at least with respect to the matters on which there is an agreement among the physicians. In that event, limits on the size of the joint venture may apply.

Statement 9 of the 1996 guidelines says that horizontal networks of all kinds will be subjected to review under the "Horizontal Merger Guidelines" to determine whether they have too much market power. These guidelines set forth a methodology for calculating market shares for competing businesses, and then set forth a methodology for determining whether a merger between two or more of the competing businesses would result in an illegal aggregation of market power.

VARIATIONS ON THE MESSENGER MODEL

This model is cumbersome and expensive to use. Therefore, physicians often wish to push the envelope of the model by adopting shortcuts that reduce the amount of time that must be invested.

Some of these variations are as follows:

- **Variation:** Same facts as above, except that the messenger "jawbones" payors that make offers that do not fall within the preauthorized fee ranges of most of the physicians and have to be relayed back to them for acceptance

[*] U.S. *v.* Health Choice of Northwest Missouri Inc., Civil Action No. 95-6171-CVSJ6, U.S.D.C. W.D. Mo., Final Judgment and Competitive Impact Statement, 9/13/95; U.S. *v.* Healthcare Partners. Inc., Civil Action No. 395-CV-01946-RNC, U.S. District Court, D. Conn., Final Judgment and Competitive Impact Statement, Filed 9/13/1995.

or rejection. The messenger tries to persuade the payor to make an offer that larger percentages of physicians have preauthorized.
    - **Legality:** According to the DOJ and FTC, this is illegal price fixing because the messenger is negotiating with the payor. The messenger cannot act as the negotiating agent of the physicians.
- **Variation:** Same facts as above, except that instead of using preauthorized fee ranges, the physicians decide that the messenger should negotiate an agreement with the payor. To avoid the bar on price fixing, the "negative option" is used—the negotiated agreement is relayed back to each physician individually for acceptance or rejection. Each physician is deemed to have accepted the offer unless the physician affirmatively states that the physician does not want to be included.
    - **Legality:** This arrangement is illegal. The messenger comes too close to negotiating an agreement on behalf of all of the physicians. Statement 9 says that use of the negative option is not necessarily enough to save an otherwise illegal arrangement. Use of the negative option may be legal if the messenger does not engage in any negotiations with the payor after presenting the price information from the physicians. In that context, the negative option may be a way to reduce the expense of this model.
- **Variation:** Same facts as above, except that 65 percent of the physicians in the community plan to be in the arrangement.
    - **Legality:** As discussed above with regard to size limits, it is uncertain whether this will be legal. The DOJ/FTC become uncomfortable when large percentages of physicians in a market are part of the same cooperative arrangement. However, when the messenger model is used, technically there is no agreement restraining trade among the physicians. The only thing that can be said with certainty is that as percentages increase, the DOJ/FTC are more likely to investigate.

NON-INTEGRATED NETWORK THAT PRESENTS AND DISCUSSES NON-FEE RELATED INFORMATION AND USES THE MESSENGER MODEL FOR FINANCIAL ARRANGEMENTS

Note: These scenarios and determinations of legality are verbatim from Frequently Asked Questions provided by the DOJ/FTC. They are presented as informational and do not constitute legal or professional advice. Antitrust laws are always decided on case specifics.

Case Scenario

The physicians in solo and small group practice in a community, which constitute 45 percent of the physicians, decide to organize a network to contract with managed-care plans. The physicians want the network for four reasons. First, they are concerned about losing patients and they want to contract with managed-care plans in order to gain patient referrals. They are willing to discount fees in return for referrals. Second, the market has become very competitive and new forms of payment have entered the market. In order to know whether offers they make will be competitive or not, the physicians want to take a survey of fee levels and other fee-related terms that are prevalent in the market. Third, they want to have influence over the medical policy of the plans, as they are concerned about the quality of care that results from the medical policies of some of the plans. Fourth, they want to have input into health plans about administrative matters that affect their practices, such as preauthorization procedures and patient grievance procedures.

To take a survey, the physicians will have a county medical society perform a survey of fees charged by physicians in the market and a survey of other matters, such as medical policy and administrative procedures used by health plans. The surveyor will get information from more than five physicians, and will aggregate the information in such a way that no one physician accounts for more than 25 percent of the content of any one statistic on an average weighted basis, and it is not possible to discern the charges of any given physician in the market. After aggregation, the messenger will share data that is at least three (3) months old with the physicians. The physicians will consider that data when deciding on discounts.

To make financial arrangements with payors, the physicians plan to have a healthcare consultant or an executive from their county medical society act as a messenger. Each physician will communicate with the messenger individually and authorize the messenger to accept contracts that fall within a fee range. The physicians will not communicate with each other about fees, and the messenger will not share the fee information with the physicians. The messenger will create two schedules. One will show the results of the survey of fees prevalent in the market. The other will show the percentage of physicians in the network willing to accept various fee levels, and will accept contract offers made by payors, that fall within the fee range of any physician or which are better. Offers lower than the fee range of any physician will be forwarded to the physician for acceptance or rejection. However, the messenger will not negotiate with or jawbone the payor.

The physicians in the network will also appoint a committee to consider the medical policies or the plans that they contract with. The committee members will solicit concerns or complaints about any contracting payor's medical policy, and they will also refer to the survey of medical policy taken by the medical society. The committee will then consider whether to recommend changes in policy to a payor. The committee may obtain information about the medical issue involved from literature, and may draft recommended protocols. If the committee decides that there should be a change, it will then present the information to the payor in question and attempt to persuade the payor to change its medical policy and adopt a protocol recommended by the committee. However, the committee will not threaten to boycott the plan if it does not adopt its recommendations, nor will it implement a boycott.

The physicians will also form a committee to consider the administrative procedures of the plan that they contract with. They will solicit concerns and complaints from the physicians in the network about administrative procedures of the plans, they will consider the survey of procedures taken by the medical society, and they will consider ways that the procedures could be improved. The committee will then meet with the health plan management to discuss its recommendations and attempt to persuade them to adopt them. However, the committee will not threaten a boycott or implement a boycott.

Legality

This arrangement is legal. The messenger model meets the criteria of Statement 9. The collective provision and discussion of medical data falls within the safety zone of Statement 4 of the 1996 guidelines, and the collection and exchange of fee information are in Statements 5 and 6. The collective discussion of administrative matters would be analyzed under the rule of reason, but meets the criteria of Statement 4.

Variation on the Arrangement

75 percent of the physicians in the community will be in the network.

Legality

According to the DOJ/FTC, the guidelines do not comment on network size for the messenger model. It appears that because there is no agreement, that it can be very large. However, once the network starts assuming other functions, such as agreeing on medical policy and administrative policy and trying to persuade plans to accept those policies, then it starts to look more like a joint venture that is subject to size limits. But, the network described may not have market power even if 75 percent of the physicians participate, because it does no more than have dialogues with plans and attempt to persuade them to adopt policies.

Variation on the Arrangement

A payor reviews the information submitted and then submits an offer that is lower than the messenger has authority to accept for more than 15 percent of the network physicians. The messenger objects, points out that the offer made is well below fee levels currently prevailing in the market, and tries to get the payor to make an offer that the messenger can accept on behalf of at least 75 percent of the physicians. The messenger points out that this level is below the average fee level prevailing in the market. To support the case for higher fees, the messenger points out that the payor can have a network right away, but the messenger will have to send lower fee offers back to the physicians for individual consideration and find out which ones may be willing to accept it, a process that could take several weeks. At the end of that time, it might be that not many would have accepted, so the payor and the messenger would have to start over again, taking still more time.

Legality

According to the DOJ/FTC, this is clearly illegal. The messenger is negotiating with the payor.

Regrettably, both the creation and the operation of some provider networks raise the risk of violating both federal and state antitrust laws. Healthcare providers have been the subject of recent antitrust enforcement agencies activities by both federal and state agencies. It is thus incumbent upon healthcare providers to analyze carefully the structure and operation of a provider network to minimize their potential antitrust risk.

QUALIFIED MANAGED CARE PLANS (QMCPs)

Another innovation in network structure that may help healthcare providers compete more effectively in the managed-care environment is the Qualified Managed Care Plan (QMCP).

First announced in 1995 by the Antitrust Division of the DOJ and the Connecticut Attorney General in consent decrees entered into with PHOs, and their IPAs, a QMCP has also been proposed as a form of resolution as recently as April 1997 in a DOJ action against a Louisiana hospital and its PHO.

A QMCP may provide physician networks with considerable flexibility in achieving compliance with the antitrust laws and may also be adaptable to many other types of provider organizations.

How the QMCP Concept Came About

The 1994 Statements of Antitrust Enforcement Policy and Analytical Principles Relating to Health Care and Antitrust (the guidelines), issued jointly by the DOJ and the FTC, outline the methodology utilized by the federal antitrust enforcers to evaluate, among other functions, compliance by a provider network with the federal antitrust laws. Physician networks that meet the requirements of the "safety zone" contained within Statement 8 of the guidelines will not be challenged, "absent extraordinary circumstances." These requirements are as follows:

1. The network, if exclusive, must include no more than 20 percent of the physicians in a specialty in the relevant geographic market (30 percent in the network is nonexclusive); and
2. The network's physicians must share substantial financial risk as a group, either through capitation or financial withholds based on preset cost-containment goals.

Simply put, a network's market share and the degree to which its members are economically integrated through shared financial risk are critical to an assessment of a network's antitrust exposure.

In many previously adjudicated antitrust cases, PHOs were alleged to have set fee schedules and negotiated collectively on behalf of their members even though those members did share substantial financial risk and/or to have contained a far greater number of physicians than permitted by the safety zone.

A QMCP may also be an appropriate vehicle for other types of provider networks (for example, networks of healthcare institutions or individual practitioners other than physicians). Because the QMCP has only been applied to physician networks, however, other types of networks must be carefully analyzed on a case-by-case basis.

Participants who share substantial financial risk through a QMCP have considerable flexibility when setting fees and negotiating with payors for network contracts, but should exercise caution to avoid allegations of illegal price fixing for services to out-of-network patients.

A provider network that operates on a non-risk-sharing basis (for example, contracts on a fee-for-service basis without withholds with some or all payors) must, in accordance with the guidelines, use a "messenger model." Under this approach, neither the network nor the messenger may negotiate collectively for the provider members with respect to fees and "other competitive terms and conditions" of managed-care contracts.

Federal antitrust enforcement officials have stated that they may release revisions to the guidelines sometime during the summer of 1997. The agencies are considering other acceptable types and levels of shared financial risk, less restrictive messenger models, and other methods by which to permit provider networks to operate with a greater degree of flexibility in this rapidly changing marketplace.

Qualified in the term QMCP means qualified only with respect to antitrust concerns. Other equally important legal issues, well beyond my scope as a paralegal, may be involved in the structuring and operating of a provider network, including Medicare/Medicaid fraud and abuse, Stark anti-self referral provisions, Medicare/Medicaid Physician Incentive Plan rules, tax exemption and corporate formation issues, corporate practice of medicine limitations, employee and ERISA issues, utilization review, licensure and liability (in Texas), and corporate governance. Always check with professional counsel with expertise in health law.

A distinction has been made between *exclusive* and *nonexclusive* physician network joint ventures. In an exclusive venture, the network's physician participants are restricted from individually contracting or affiliating with other network joint ventures or health plans. In a nonexclusive venture, the physician participants are available to affiliate with other networks or contract individually with health plans.

Factors that the agencies will examine on the question of nonexclusivity include the following:

1. That viable competing networks or managed-care plans with adequate physician participation currently exist in the market
2. That physicians in the network actually individually participate in, or contract with, other networks or managed-care plans, or there is other evidence of their willingness and incentive to do so
3. That physicians in the network earn substantial revenue from other networks or through individual contracts with managed-care plans
4. The absence of any indications of significant de-participation from other networks or managed-care plans in the market
5. The absence of any indications of coordination among the physicians in the network regarding price or other competitively significant terms of participation in other networks or managed-care plans

If contract provisions significantly restrict the ability or willingness of network physicians to join other networks or contract individually with managed-care plans, the agencies will consider the network to be exclusive for the purposes of antitrust "safety zones."

Substantial Financial Risk Must Be Shared

To qualify for either the exclusive or nonexclusive physician network joint venture antitrust "safety zone," the participants in a physician network joint venture must share substantial financial risk in providing all the services that are jointly priced through the network. Examples of some types of arrangements through which participants in a physician network joint venture can share substantial financial risk include the following:

1. Agreement by the venture to provide services to a health plan at a "capitated" rate
2. Agreement by the venture to provide designated services or classes of services to a health plan for a predetermined percentage of premium or revenue from the plan
3. Use by the venture of significant financial incentives for its physician participants, as a group, to achieve specified cost-containment goals. For example, two methods by which the venture can accomplish this are
   a. Withholding from all physician participants in the network a substantial amount of the compensation due to them, with distribution of that amount to the physician participants based on group performance in meeting the cost-containment goals of the network as a whole, or
   b. Establishing overall cost or utilization targets for the network as a whole, with the network's physician participants subject to subsequent substantial financial rewards or penalties based on group performance in meeting the targets
4. Agreement by the venture to provide a complex or extended course of treatment that requires the substantial coordination of care by physicians in different specialties offering a complementary mix of services, for a fixed, predetermined payment, where the costs of that course of treatment for any individual patient can vary greatly due to the individual patient's condition, the choice, complexity, length of treatment, or other factors

Inasmuch as new risk sharing arrangements are constantly emerging in this changing healthcare marketplace, those creating physician network joint ventures can also take advantage of the DOJ's expedited business review procedure or the FTC's advisory opinion procedure.

Agency Analysis of Physician Network Joint Ventures that Fall Outside These Antitrust "Safety Zones"

Physician network joint ventures that fall outside the antitrust "safety zones" do not necessarily raise substantial antitrust concerns. Likewise, those joint ventures that do not involve the sharing of substantial financial risk may also be viewed as noncompetitive. It is not the agencies' intent to treat physician networks either more strictly or more leniently than joint ventures in all other industries. Instead, the agencies' goal is to ensure a competitive marketplace in which consumers will have the benefit of high-quality, cost-effective healthcare and a wide range of choices, including new provider-controlled networks that expand consumer choice and increase competition.

Therefore, where competitors economically integrate in a joint venture, such agreements, if reasonably necessary to accomplish the pro-competitive benefits of the integration, are analyzed under the rule of reason analysis. The agencies will basically apply the following type of evaluation (Note, however, that the agencies' ultimate conclusion is based upon a more comprehensive analysis.):

1. They will define the relevant market.
2. They will evaluate the competitive effects of the physician joint venture.
3. They will evaluate the impact of pro-competitive efficiencies that result from this joint venture.
4. They will evaluate collateral agreements or conditions that unreasonably restrict competition and are unlikely to contribute significantly to the legitimate purpose of the physician network joint venture.

In this policy statement, the agencies provide seven examples of physician joint ventures and subject them to rule of reason analysis. As a paralegal, I am not licensed to be able to provide specific advice as to whether your integrated network joint ventures are in compliance with federal antitrust law and these guidelines. Every joint venture is unique, and the application of the antitrust laws can be quite complex. Retaining legal counsel is essential if you are considering entering into or creating an integrated network joint venture.

Statement 9: Multi-Provider Networks

Healthcare providers are forming a wide range of new relationships and affiliations, including networks among otherwise competing providers, as well as networks of providers offering complementary or unrelated services. Because multi-provider networks involve a large variety of structures and relationships among many different types of healthcare providers, and new arrangements are continually developing, the agencies are unable to establish a meaningful antitrust "safety zone" for any of these arrangements.

As stated previously, the antitrust laws condemn as per se illegal naked arrangements among competitors that fix prices or allocate markets. However, a "rule of reason" analysis will be applied and the network will not be viewed as per se illegal if the providers' integration through the network is likely to produce significant efficiencies that benefit consumers, and any price agreements (or other agreements that would otherwise be per se illegal) by the network providers are reasonably necessary to realize those efficiencies.

Shared Substantial Financial Risk

In some multi-provider networks, significant efficiencies may be achieved through agreement by the competing providers to share substantial financial risk for the services provided through the network; in such cases, the setting of price would be integral to the network's use of such an arrangement and therefore would warrant evaluation under "rule of reason." Examples of substantial financial risks can be found in Section 8 of this fact sheet. Organizers of multi-provider networks who are uncertain whether their proposed arrangements constitute substantial financial risk sharing can also take advantage of the DOJ's expedited business review procedure or the FTC's advisory opinion procedure.

No Sharing of Financial Risk

Multi-provider networks that do not involve the sharing of substantial financial risk may also be sufficiently integrated to demonstrate that the venture is likely to produce significant efficiencies. These arrangements will also be analyzed under "rule of reason."

Rule of Reason Analysis

In rule of reason analysis of multi-provider networks, the agencies will basically apply the following type of evaluation (Note, however, that the agencies' ultimate conclusion is based upon a more comprehensive analysis.):

1. They will evaluate the competitive effects of multi-provider networks in each of the relevant markets in which these networks operate or have substantial impact.

   The relevant geographic market for each relevant product market affected by the multi-provider network will be determined through a fact-specific analysis that focuses on the location of reasonable alternatives. Therefore, the relevant geographic market may be broader for some product markets than for others.
2. They will examine the competitive effects of the network, both the potential "horizontal" and "vertical" effects of the arrangement, and its exclusion of particular providers.
3. The agencies will review the balance of any potential anti-competitive effects of the multi-provider network against potential efficiencies associated with its formation and operation. The greater the network's likely anti-competitive effects, the greater must be the network's likely efficiencies.

In conducting a rule of reason analysis, the agencies rely upon a wide variety of data and information, including information supplied by the participants in the multi-provider network, purchasers, providers, consumers, and others familiar with the market in question. It is not simply a question of polling those who support and oppose the formation of the network. Some common arrangements include selective contracting and messenger models.

Selective Contracting

Exclusion of particular providers may be a method by which networks limit provider plans in an effort to achieve quality and cost-containment goals, and thus enhance their ability to compete against other networks. Selective contracting may also be pro-competitive by giving nonparticipant providers an incentive to form competing networks. A rule of reason analysis will again be applied. In examining exclusive arrangements, the agencies will examine the degree to which the arrangements may limit the ability of other networks or health plans to compete in the market. The focus of the analysis is not on whether a particular provider has been harmed by the exclusion or referral policies, but rather whether the conduct reduces competition among providers in the market and thereby harms consumers.

Messenger Models

Some networks that are not substantially integrated use a variety of "messenger model" arrangements to facilitate non-integrated network contracting and avoid price-fixing agreements among competing network providers. Arrangements that are designed simply to minimize the costs associated with the contracting process that do not result in a collective determination by the competing network's providers on prices or price-related terms, are not, per se, illegal price fixing.

The key issue in any messenger model arrangement is whether the arrangement creates or facilitates an agreement among competitors on prices or price-related terms. Determining whether there is such an agreement is a question of fact in each case. The agencies will examine whether the agent facilitates collective decision making by network providers, rather than independent, unilateral decisions. In particular, the agencies will examine whether the agent coordinates the providers' responses to a particular proposal, disseminates to network providers the views or intentions of other network providers as to the proposal, expresses an opinion on the terms offered, collectively negotiates for the providers, or decides whether or not to convey an offer based on the agent's judgment about the attractiveness of the prices or price-related terms. If the agent engages in such activities, the arrangement may amount to a per se illegal price-fixing agreement.

In this policy statement, the agencies provide four examples of multi-provider network joint ventures, and subject them to antitrust analysis. As a paralegal, I cannot provide specific legal advice as to whether your multi-provider networks are in compliance with federal antitrust law and these guidelines. Every multi-provider network joint venture is unique, and the application of the antitrust laws can be quite complex. Retaining legal counsel is essential if you are considering entering into, participating in, or creating a multi-provider network joint venture.

Useful Addresses and Telephone Numbers

To contact the Antitrust Division regarding business review letters, call or write the agency at

Legal Procedure Unit
Antitrust Division
U.S. Department of Justice
325 7th Street, N.W.
Suite 215
Washington, D.C. 20530
Telephone number: (202) 514-2481

To contact the Federal Trade Commission (FTC) regarding advisory opinions, call or write the agency at

Health Care Division
Bureau of Competition
Federal Trade Commission
Washington, D.C. 20580
Telephone number: (202) 326-2756

13 Business Plan Development

As an IPA (Independent Practice Association), PHO (Physician Hospital Organization), or MSO (Management Services Organization), it will be necessary to state your business goals, objectives, and projections in words for others to be able to read and evaluate.

I have found that many Business Plan software applications simply do not "fit" well for this type of business, as they do not take you through the necessary thought processes involved in such a complicated organization. They do work really well for donut shops though!

Before you begin, here are some common errors you may wish to avoid as you endeavor this "literary work":

1. Be absolutely clear on the purpose for which you are writing the business plan: The planning matters most, not how fancy your document is and its eye appeal. Provider organizations need to engage in planning the business because planning begets management. Planning is a process of setting goals and establishing specific measures of progress, and then tracking your progress and following up with course corrections. The plan itself is just the first step; it is reviewed and revised often. Do not even print it unless you absolutely have to. Leave it on a shared digital network instead. Google Apps makes this shared digital network available for up to ten users, which is, at present, free of charge. If you establish a nonprofit, I believe the allowance is up to fifty users before you have to pay for the service.
2. Do not attempt to write the whole plan in one protracted editorial session (or have a consultant do it for you); do it in pieces and steps. The plan is a set of connected modules, like building blocks. The reason I created this book in chunks is so you can start with any focal point and just get busy. First, complete the parts that interest you most, or the parts that provide the most immediate benefit. That might be strategy, concepts, target markets, business offerings, projections, mantra, vision, whatever ... just start!
3. Do not ever consider your plan "done." It is a dynamic document. I consulted for a physician recently who told me his accountant wrote the plan; it got used to get financing, and he did not even know where it was, had not seen it in two years. No wonder his practice was in trouble. If your plan is done, then your business is done. That most recent version is just a snapshot

of what the plan was then. It should always be alive and changing to reflect changing assumptions.

4. Do not hide the plan from the Board or the Steering Committee members. Do protect parts of it as proprietary and competitive. It is a management tool. Use common sense about what you share with everybody on your team, keeping some information, such as competitive analysis, strategic research, business intelligence, and individual salaries, confidential. But do share the goals and measurements, using the planning to build team spirit and peer collaboration. That does not mean sharing the plan with outsiders, except when you have to, such as when you are seeking capital. In most cases, you will not need to share the entire plan with all physician investors, just the first 3 to 5 pages, which is generally an Executive Summary and high-level financials.
5. Do not confuse cash with profits. This happens a lot in small medical groups. "There is cash left in the account this month, so we made a profit." Actually, there is a huge difference between the two. Waiting for health plans to pay can cripple your financial situation without affecting your profits. Loading your inventory absorbs money without changing profits. Profits are an accounting concept; cash is money in the bank. You do not pay your bills with profits.
6. Avoid distractions and diluting your priorities. Stay focused. Hone your plan to feature three or four priorities for focus and power. People can understand three or four main points. A plan that lists twenty priorities does not really have any. If you have more, see if they cannot be logically grouped and pared down in number.
7. Overvaluing the business idea. Egos get in the way with entrepreneurs. It is our nature. What gives an idea value is not the idea itself but the business that is built on it. You cannot conceptualize value; you have to produce it. In healthcare, services being rendered and customers paying their bills converts an idea into a business. The rubber meets the road when you are forced to either write a business plan that shows you building a business around that great idea, or forget it.
8. Fudging the details in the first 12 months. If you have shareholders (physicians who capitalized the IPA or MSO), they are owed a duty of honesty and accountability. By details, I mean your financials, milestones, responsibilities, and deadlines. Cash flow is most important, but you also need lots of details when it comes to assigning tasks to people, setting dates, and specifying what is supposed to happen and who is supposed to make it happen. These details really matter. A business plan is wasted without them or if the business plan is not really adhered to after it is written.
9. Sweating the details for the later years. This is a business plan, not an accounting plan. How can you project monthly cash flow three years from now when your sales forecast is so uncertain? (Thank you, Capitol Hill!) Sure, you can plan in five-, ten-, or even twenty-year horizons in the major conceptual text, but you cannot plan in monthly detail past the first year. Nobody expects it, and nobody believes it.

10. Making absurd forecasts. Nobody believes absurdly overestimated sales projections. If you have them, you will need to answer why you have them in terms that business school people will be satisfied. Forecasting unusually high profitability usually means you may not have a realistic understanding of expenses.

MODEL BUSINESS PLAN

Below I layout a general template that you can use as sample summary report for a fictitious IPA, organized as an LLC. As you read through the next few pages, think of how you might fill in the blanks for your organization.

As a first step, decide what to put into your business plan and a direction. Break each part into a standard essay format with a thesis statement and introductory paragraph or two, a few supporting paragraphs, and a conclusion. An example that is most popular with start-ups follows:

- Table of Contents
- Executive Summary Financial Plan
- Assumptions Financial Statements
- Capital Requirements Use of Funds
- Exit/Payback Strategy Conclusion

For the sake of demonstration, I have abbreviated the business plan to simply give you an idea of how the document flows. Normally, business plans are very formal documents with much elaboration on the areas I have highlighted. What I have found out is that your investor or banker wants to see a well-designed plan, with a viable rationale and lots of substance instead of "fluff."

EXECUTIVE SUMMARY

This section provides several parts—namely, Company Direction, Company Overview (background, objectives, and capital requirements), Management Team, Service Strategy, Market Analysis, Customer Profile, Marketing Plan, Marketing Strategy, Advertising and Promotion, Public Relations, Financial Plan, and Conclusion.

Following is a sample of an Executive Summary.

Company Direction

In 1997, [business name] was founded to form an organization of healthcare providers that could

1. Exercise some contracting clout with managed-care payors,
2. Share financial risk,
3. Preserve clinical autonomy and access to patients, and
4. Promote professional camaraderie between providers in the community while delivering quality healthcare at an affordable price.

Overall, our company can be characterized as a high-profile, aggressive, healthcare provider organization known for its quality care and business leadership in the community.

Company Overview

Background: For many years, providers have contracted directly with insurance plans, HMOs, PPOs, and others, often at a disadvantage, without the leverage needed to negotiate from a position of strength.

The condition of the industry today is such that managed healthcare plans are deferring the network development activities to organized groups of providers ready to share risk with them.

The legal form of [business name] is Limited Liability Company. We chose the Limited Liability Company form because of the inherent flexibility of size and tax advantages.

[Business name]'s business headquarters is located at [your address].

Revenue projected for fiscal year [current fiscal year] without external funding is expected to be [current internal revenue dollars]. Annual growth is projected to be [annual growth percent] through [growth ending year].

Now, [business name] is at a point where [future needs/wants].

Objectives

Based on our projected revenues for the current fiscal year and our projected annual growth, we feel that within [time span] years, [business name] will be in a suitable position for [future company position]. Our objective, at this time, is to propel the company into a prominent market position.

Capital Requirements

According to the opportunities and requirements for [business name] described in this business plan, and based on what we feel are sound business assumptions, our [capital time frame] capital requirements are for [capital required dollars].

To accomplish this goal, we have developed a comprehensive plan to intensify and accelerate our marketing and sales activities, services expansion, and customer service. To implement our plans, we require an estimated total of [five-year loan dollars] in financing over the next five years for the following purposes:

- [fund use]

We require additional investments of [additional capital dollars and time frame] in order to increase our production capacities to meet market demand.

Management Team

Our management team consists of six men and women:

- Dr. Harry Jones, M.D., President and CEO
- Dr. Susan James, Vice President
- Punjit Ala, M.D., Secretary

- Sandra Allan, D.O., Treasurer
- Raul Martinez, M.D., Medical Director
- Mohammed Khaldi, M.D., Director of Operations

Their backgrounds consist of more than [overall experience] years of experience. Our CEO and advisory staff have a total of [corporate development experience] years of managed-care experience and IPA participation.

Additionally, our outside management advisors provide tremendous support for management decisions and creativity:

- [outside support name and title]

Service Strategy

Current Service Medical and Surgical services to managed-care participants, Medicare and Medicaid patients, and others with and without private health insurance coverage.

Caring for our patients is our primary focus.

Market Analysis

Market Definition

The managed-care delivery system market is growing at a rapid rate. The market for HMO growth amounted to [market dollars] in [market year]. This represents [market growth percent] growth over the past [previous market years] years.

According to market research and industry sources, the overall [market type] market for the [industry type] industry is projected to be [market dollars for industry] by the end of [market industry year]. The area of greatest growth in the managed-care market is in the area of provider risk assumption and capitation. Currently, the market is shared by [market participants], with [top competitor] considered the market leader.

Customer Profile

[Business name]'s target market includes managed-care insurance and health plans, HMOs, PPOs, EPOs, healthcare purchasing coalitions and alliances, ERISA (Employee Retirement Income Security Act) self-funded health plans, and other emerging groups. The most typical customers for our services are managed healthcare plans, and those who currently use our product for quality healthcare delivery and cost containment.

A partial list of actual customers includes

- [customer list]

Competition

Companies that compete in this market are [key competitors]. All companies mentioned charge competitive prices as follows:

- [competitor products and prices]

Key factors that have resulted in the present competitive position in this industry are

- [key competitive factors]

In all comparisons, [business name] provides superior performance compared to competitive organizations because of our well-developed organizational infrastructure and our ability to share substantial risk among all members. In most cases, the number of differences is substantial. A complete technical comparison is available.

[Business name]'s network can participate in virtually all situations involved with managed healthcare delivery. The ability to be flexible to the needs of the customer with full capability on utilization management, quality improvement, and severity adjusted, acuity indexed outcomes is unique to [business name]'s network. The ability to demonstrate this using our own data is unique to our network, and our research indicates its performance is superior to anything else in our market today.

Risk

The top business risks that [business name] faces are

- [top business risks]

The economic risks affecting [business name] are [economic risks].

Marketing Plan

Responses from customers indicate that our network is enjoying an excellent reputation, and we fully intend to continue this trend. Inquiries from prospective customers suggest that there is considerable demand for it.

[Business name]'s marketing strategy is to enhance, promote, and support the fact that our products [product marketing strategy].

Marketing Strategy

Because of the special market characteristics [special market characteristics], our strategy includes [sales strategy].

Advertising and Promotion

[Business name]'s overall advertising and promotional objectives are to position [business name] as the leader in the market.

We will develop an advertising campaign built around [ad campaign message], beginning with a "who we are" statement and supporting it with ads that reinforce this message. Additionally, we will develop a consistent reach and frequency throughout the year. In addition to standard advertising practices, we will gain considerable recognition through [other advertising practices].

For the next year, advertising and promotion will require [advertising money needs]. On an ongoing basis, we will budget our advertising investment as 3 percent of gross revenues.

Public Relations

During [publicity strategy year], [business name] will focus on the following publicity strategies:

- [top publicity strategies]

We will track, wherever possible, the incremental revenue generated from our advertising, promotion, and publicity efforts. We anticipate at least [publicity sales dollars] of sales will be generated directly from our publicity, and possibly additional [indirect dollar increase in sales] of indirect increase in sales throughout our various channels.

Financial Plan

The financial projections indicate that profit will be achievable in two (2) years. The increase in profits generated by this investment, specifically [expected increase description], will allow us to have the funds to develop a Management Services Organization (MSO) in the next eighteen (18) months.

Conclusion

[Business name] enjoys an established track-record of excellence with our customers. Their expressions of satisfaction and encouragement are numerous, and we intend to continue our advances and growth in the marketplace with more unique and effective products.

[Executive Summary Concluding Remarks]

Financial Plan

Here is a great opportunity to enlist the assistance of the network's finance committee. A PHO may have an easier time if the finance committee and the hospital's CFO participate in the development of these documents, but IPAs that have some participant interested in finance and numbers usually do very well with this section.

First we start with a section of assumptions about when the network will be up and running, and how long it will take to reach profitability and projected market penetration. Following that is a section that actually projects some numbers for total revenue, gross profit, and income from operations, followed by income ratios, gross profit analysis, and a budget with income statements, balance sheets, and cash flow statements.

The reader is going to be curious about how the money will be used and how you intend to service your debt and offer exit strategies to those who may leave the organization. Be sure to include these statements and be honest and forthright. If you do not have an answer, seek professional assistance for these answers from an experienced financial consultant.

Assumptions

Our financial projections are based on the following assumptions:

A working network will be available by [working prototype date].

Initial market penetration is anticipated to be [market penetration dollars] at a margin of [market penetration margin percent]. This is expected to increase to [year 1 increase market penetration dollars] by the end of Year 1 and to [year 5 increase market penetration dollars] by the end of Year 5.

Gross Profit Analysis

The Gross Profit. Analysis statements included in our supporting documents show monthly sales revenue, cost of goods sold, and gross profit values for each of our service lines for the first year.

Financial Statements Primary Income—Related Items

Year	Fiscal Year 1	Fiscal Year 2	Fiscal Year 3	Fiscal Year 4	Fiscal Year 5
Total revenue					
Growth %					
Gross profit					
Growth %					
Income from operations					
Growth %					
Net income after taxes					
Growth %					

Income Ratios

	Year 1	Year 2	Year 3	Year 4	Year 5
Gross profit margin					
Operating income margin					
Net profit margin					
Return on equity					

Budget—Income Statements

There are three income statements included in our supporting documents:

- Year 1 by month
- Years 2 and 3 by quarter
- Years 1 through 5 Balance Sheets

Balance Sheets

There are three balance sheets included in our supporting documents:

- Year 1 by month
- Years 2 and 3 by quarter
- Years 1 through 5

Cash Flows Statements

There are three statements of changes in financial position (Cash Flows Statements) included in our supporting documents:

- Year 1 by month
- Years 2 and 3 by quarter
- Years 1 through 5

Break-Even Analysis

The break-even analysis included in our supporting documents indicates that the break-even point will be reached in [break-even month and year].

Revenue is projected to be [dollars above break-even] above break-even in [revenue above break-even month and year].

The contribution margin for the first year is [amount] percent, which represents a break-even sales volume of [$ amount], and a sales volume above break-even of [$ amount].

Capital Requirements

The [capital time frame] capital required is [capital required dollars]. We require additional investments of [additional capital dollars and time frame] in order to increase our production capacities to meet market demand.

After analyzing our working capital, we estimate our operating working capital requirements as [year 1 working capital], [year 2 working capital], [year 3 working capital], [year 4 working capital], and [year 5 working capital] for years 1 through 5, respectively. We will need to borrow [working capital finance dollars] to finance working capital for a period of [working capital finance time frame]. The remainder will be financed through cash from operations.

In order to purchase [purchase additions], an estimated total of [5-year loan dollars] in financing is required for the next five-year period. The annual requirements for each year are estimated as [year l loan dollars], [year 2 loan dollars], [year 3 loan dollars], [year 4 loan dollars], and [year 5 loan dollars], respectively.

The level of safety is [safety level] for this [industry or investment type]. Our confidence in achieving the attached financial projections within [achievement percent] is [confidence level]. In addition to the operation of the business, additional protection is provided by [protector] as collateral. Were the situation to arise where the collateral was needed, the realizable value of the collateral would be [collateral value dollars], which reduces the amount at risk to [at-risk dollars]. With a projected return of [projected return dollars], this represents a return of [at-risk return percent] of the amount at risk.

Use of Funds

The funding proceeds will be used to [fund use].

Exit/Payback Strategy

The financial projections indicate that exit of [investor name] will be achievable in [years investor exit] years. The exit settlement will be in the form of [investor exit strategy]. The increase in profits generated by this investment, specifically [expected increase description], will allow us to have the funds to repay the loan in [payback time frame].

Conclusion

[Business name] enjoys an established track-record of excellence with our customers. Their expressions of satisfaction and encouragement are numerous, and we intend to continue our advances and growth in the marketplace with more unique and effective products.

Based on the attached financial projections, we believe that this venture represents a sound business investment.

In order to [business progression], we are requesting a [financing type] of [financing amount] by [financing date].

[end sample document]

A business plan need not be a hindrance, so try not to become so bogged down in specifics that you cannot be flexible as a company to meet market changes without starting the plan over from scratch. Some additional information sources are included below in case you want to do more research in this area.

KEEPING YOUR INFANT BUSINESS COMPETITIVE: NONDISCLOSURE AGREEMENTS

Following the design of your business plan, you will need to protect your interests as best as possible. You can start with a non-disclosure statement for those who you allow to read the plan, so that you have not given them the cookbook to build a competitive organization as a clone to yours. The following non-disclosure statement model may help you.

Sample Nondisclosure Agreement

The undersigned acknowledges that [business name] has furnished to the undersigned potential Investor ("Investor") certain proprietary data ("Confidential Information") relating to the business affairs and operations of [business name] for study and evaluation by Investor for possibly investing in [business name].

It is acknowledged by Investor that the information provided by [business name] is confidential; therefore, Investor agrees not to disclose it and not to disclose that any discussions or contracts with [business name] have occurred or are intended, other than as provided for in the following paragraph.

It is acknowledged by Investor that information to be furnished is in all respects confidential in nature, other than information that is in the public domain through other means, and that any disclosure or use of same by Investor, except as provided in this agreement, may cause serious harm or damage to [business name] and its owners and officers. Therefore, Investor agrees that Investor will not use the information furnished for any purpose other than as stated above, and agrees that Investor will not either directly or indirectly by agent, employee, or representative, disclose this information, either in whole or in part, to any third party, provided, however, that any information furnished may be disclosed only to those directors, officers, and employees of Investor and to Investor's advisors or their representatives who need such information for the purpose of evaluating any possible transaction (it being understood that those directors, officers, employees, advisors, and representatives shall be informed by Investor of the confidential nature of such information and shall be directed by Investor to treat such information confidentially), or for any disclosure of information may be made to which [business name] consents in writing. At the close of negotiations, Investor will return to [business name] all records, reports, documents, and memoranda furnished and will not make or retain any copy thereof.

Signature ______________________________ Date ______________
Name (typed or printed)
[Addressee Company]
[end sample document]
This is a business plan. It does not imply an offering of securities.

Remember: "To Fail to Plan is to Plan to Fail"

14 Guidance for the IT Committee

In the face of healthcare reform, regardless of its final political outcome, one thing is for sure and will never be unraveled. The advancements in healthcare information technology (HIT) will not slow down or be abandoned.

In the United States, the Patient Protection and Affordable Care Act (HR 3590) and Health Care and Education Reconciliation Act (HR 4872) were passed to address several problems with healthcare in the United States.

Currently, the situation we all face includes these four harsh realities:

1. We are currently spending 17 percent of our Gross Domestic Product on healthcare, yet we have worse population health outcomes than many other industrialized societies spending half as much.
2. Healthcare costs are rising faster than inflation.
3. We have significant variation in practice patterns that is neither explained by patient comorbidities nor justified by comparative effectiveness evidence.
4. We want to expand access to health insurance to 95 percent of the population, lower our spending growth rate, and incentivize delivery system change.

Healthcare payment reform transforms the entire healthcare reimbursement system from fee-for-service to value-based payment—paying for good outcomes rather than quantity of care. Pilot projects will test new payment methods and delivery models. Successful innovations will be widely implemented. This will not happen in just Medicare, Medicaid, and ACOs; it will be universal across the entire set of healthcare reimbursement channels in very short order.

Health insurance reform expands coverage, makes features and costs of plans transparent, and removes the barriers to enrollment created by preexisting condition considerations. These three realities will have positive and negative effects in a world of capitated or bundled reimbursement for healthcare services.

- If capitated, service equals expense instead of revenue.
- If bundled and the bundled elements of what goes into a "case" or "episode of care" are not clearly defined, the ones receiving the reimbursement are bound for bankruptcy if they are not adequately reinsured for the risk of "scope creep."

In both situations, without adequate information systems and technology to help manage data collection, create analytical reports, and monitor reimbursement against costs, the system is perfunctorily doomed. It becomes simply a matter of how long it takes to declare the failure and mark the time and date of expiration.

As healthcare providers, you have very little influence over both healthcare insurance reform and healthcare payment reform. You have ultimate influence over **healthcare delivery reform**.

The Provider Organization HIT Committee will focus on the implications of payment reform that will lead to delivery system reform.

Medicare initiatives include

- Medicare shared savings program, including Accountable Care Organizations (ACOs)
- National pilot program on payment bundling
- Independence at Home demonstration program
- Hospital re-admissions reduction program
- Community-based care transitions program
- Extension of gain-sharing demonstration
- Physician Quality Reporting Initiative (PQRI)
- Group Practice Reporting Option (GPRO)
- Hospital Consumer Assessment of Health Plans Survey (or Hospital CAHPS®) Surveys
- Value innovation quality scoring transparency and publicity

Medicaid initiatives include

- Health Homes for the Chronically Ill
- Medicaid Community First Choice Option
- Home- and Community-Based Services State Plan Option
- Hospital care integration
- Global capitation payment for safety net hospitals
- Pediatric ACOs

Managed-care initiatives include

- A return to capitation payment methods
- Bundled case rates in fee-for-service settings
- Increased steerage to high-performing providers
- Direct contracting with self-funded employers, unions, TPAs, and purchasing coalitions
- Health travel incentive options steering to narrow networks
- Health Insurance Exchanges (HIEs)
- Value innovation quality scoring transparency and publicity

So, as the HIT Committee of the new provider organization, your role includes making decisions about the brand of computer to purchase for the office that will manage the network, and maybe even to determine which tablet system will be "cool" to have as officers. And it will not even be as simple as ensuring that the network has HIPAA Business Associate Addenda for everyone on file. Those are all your role and responsibility too. But that is child's play.

In anatomy-speak, you are the "Circle of Willis" of the provider organization. For readers who are not that fluent in anatomy and physiology, the Circle of Willis is a unique vascular structure in the body that provides an extended safety net of redundancy for the brain's blood supply. No blood, no oxygen to the brain. Everybody knows what that means. But it does not stop there. The analogy continues: the Circle of Willis, because of its complexity, varies anatomically among individuals and is a common site for congenital (present at birth) vascular anomalies (malformations of the blood vessels). Think of your subcommittees as the blood vessels. Now let me take you to a place you probably did not expect to go in this chapter.

For example, since its emergence more than twenty years ago, business intelligence (BI) has never experienced as much rapid change as it has in the recent months. BI tools have become less expensive while becoming more visual, interactive, analytical, mobile, and scalable—which allows end users to customize dashboards with little IT technician or support desk involvement.

Many BI tools now incorporate in-memory databases or mid-tier caches to improve performance and interactive visualizations to enhance the user experience. These new technologies such as in-memory analytics will allow network providers and administrators to utilize data for more insight into their organization, and customers and vendors are taking advantage of that. The committee will need to become fluent on BI technologies to work smarter with existing data that can be gathered through the provider organization without violating HIPAA (Health Insurance Portability and Accountability Act) regulations.

You will have to learn how BI tools will balance top-down metrics reporting with bottom-up ad hoc analysis in a user-friendly manner that delivers fast performance while securing information consistency.

There will be two user groups of the BI: those who use the reports for metrics monitoring, analyzing anomalies and drilling to details, and those who are analysts who will explore data, model it, examine the source integrity, and publish outcomes. Your job will be to plan how to accommodate each.

The two groups should be coordinated by or at least interact with the IT committee to provide alignment and integration.

The casual users will manage corporate objectives and strategies. They will use predefined metrics and nonvolatile data from a data warehouse. This will help bring alignment to the organization, together with consistency, scalability, and security. As your needs grow as an organization, this group will act as the pilots and change agents. The challenges you will have to manage as the HIT steering committee chair will be that it will be difficult to recruit good members for this team. The members that you start out with on day 1 may not be the members who are still there on day 366. You will likely deal with politically charged arguments, especially if the committee does have adequate representation from specialists, primary care, and hospital-based physicians and administrators. An imbalance can spell disaster for the potential viability of the group. This top-down-oriented group will share an ilk toward being driven by the business intelligence and runs the risk of becoming complacent with how data magically gets to them each month so they can run their reports. If not politically velveteen when interacting with the power users, there will be clashes as to whether or not the reports beget analysis or analysis begets reports.

The power users will want to be able to perform ad hoc queries, and dig into volatile data and expect it to be fresh, not stale, and easily changeable. The group will be involved in developing processes and projects such as credentialing, privileging, and clinical outcomes measurement campaigns, and will be oriented very bottom-up, and share an ilk toward analytics intelligence.

Both teams will appreciate it if you choose a system that enables them to manipulate data and create self-service dashboards for use by specialists, primary care, hospital-based specialists, and administration without having to wait for the IT technician, whose most overused word in his or her vocabulary is "move" as they take over the mouse and the requestor's seat. They will appreciate other features such as

- Interactivity
- Visually organized
- Flexibility
- Analytical
- Predictive
- Collaborative
- Mobile application capable of access from anywhere
- Fast running (dictated by file architecture and hosting platform)
- Quick to deploy
- Interoperable from any data source, such as different EHRs
- Scalable
- Maintainable by users, rather than always waiting for technicians to deliver
- Manageable (role based, security profiles, audit trails, etc.)
- Comprehensive
- Cloud-based portability

Your consumers will be process oriented and will have a hierarchy that you will need to keep top of mind. As seen in Figure 14.1, casual users will want to be able to view, navigate, and modify data, while power users will want to modify, explore, and model the data. As the committee, you will be the "mom" between the sibling rivalry. Ensure there is adequate balance and consensus on the committee or you will end up playing the "go ask your mom," "go ask your dad" game, or worse yet, "Mom said we could" or "Dad said we could."

As for consultants, you will need one with a Certified Information Systems Security Professional (CISSP) credential. These professionals are certified by the International Information Systems Security Certification Consortium (ISC)². The CISSP was the first information security credential accredited by the American National Standards Institute (ANSI) ISO/IEC Standard 17024:2003 accreditation, and, as such, has led industry acceptance of this global standard and its requirements. The CISSP curriculum covers subject matter in a variety of Information Security topics based on a common body of knowledge founded upon a fundamental triad that will be important to your committee—that of confidentiality, data integrity, and availability.

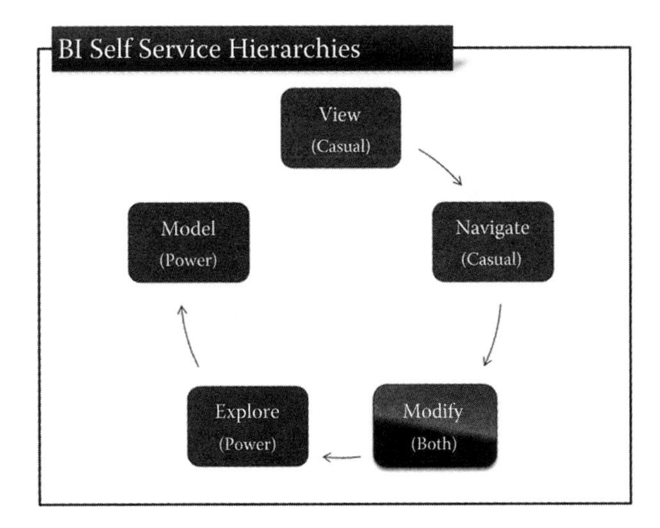

FIGURE 14.1 BI self-service hierarchies.

The right consultant will get you started in the right direction toward compliance. We have one here at Mercury for our operations who has been a real comfort. Just knowing that we have one authoritative expert as the go-to person for compliance and data security has reduced the "he said/she said" rumors and inaccurate interpretations of new standards and regulations handed down.

The other consultant you may need will be a database administrator or systems administrator.

Beyond such fundamentals as HIPAA compliance, billing, and interoperability, you will be responsible for the overall backbone of the organization's ability to operate and do business.

You will work with committee members, other committees, administrators and focus teams or subcommittees, and your consultants to address:

- Access Control:
  - Categories and Controls
  - Control Threats and Countermeasures
- Telecommunications and Network Security:
  - Network Security Concepts and Risks
  - Business Goals and Network Security
- Information Security Governance and Risk Management:
  - Policies, Standards, Guidelines, and Procedures
  - Risk Management Tools and Practices
  - Planning and Organization
- Software Development Security:
  - Software-Based Controls
  - Systems Development Lifecycle and Principles
- Cryptography:
  - Basic Concepts and Algorithms

    - Cryptography Standards and Algorithms
    - Signatures and Certification
    - Cryptanalysis
- Security Architecture and Design:
    - Principles and Benefits
    - Trusted Systems and Computing Base
    - System and Enterprise Architecture
- Operations Security:
    - Media, Backups, and Change Control Management
    - Controls Categories
- Business Continuity and Disaster Recovery Planning:
    - Response and Recovery Plans
    - Restoration Activities
- Legal, Regulations, Investigations, and Compliance:
    - Major Legal Systems
    - Common and Civil Law
    - Regulations, Laws, and Information Security
- Physical (Environmental) Security:
    - Layered Physical Defense and Entry Points
    - Site Location Principles

Remember to budget and capitalize adequately for these functions and this professional guidance and to vet consultants with healthcare BI and healthcare analytics experience. Without the functions involved here, the rest of the committees will be at a standstill—kind of like a stroke, from which the organization may never recover.

There are a few other application-level items you will be responsible for, so break up those interested in serving on this committee into subcommittees based on interests:

Contracting and reimbursement: You will need a contracts database that is navigable. It can be a flat file (spreadsheet) or relational (database); regardless, your administrator and contracting expert will be well-served with some tool to manage contracts and monitor payer compliance, timeliness, denials, appeals, and profitability through the use of a payer report card. The database should also have financial modeling capability.

Communications with the provider organization members: Think simple, like Google Apps. Powerful, secure, cheap. If you are not familiar with Google Apps, take a tour at <http://www.google.com/apps/intl/en/business/features.html>. Currently, more than four million businesses run Google Apps.

With Google Apps you can have inexpensive group e-mail, workgroup document sharing, your contracting spreadsheets for payer report cards, contacts management, mobile instant messaging, a Google VOIP* phone number, group calendars, secure encryption, spam filtering, e-mail routing and management of multiple e-mail addresses, and more. With their Cloud Connect® product, you can share MS Office® or Google Docs® for simultaneous editing for Word®, PowerPoint®, and Excel®

---

* Voice Over Internet Protocol.

files, no document or paragraph locking, have Google Docs sharing URLs for each Microsoft Office® file for workgroups, track revisions with an automatic revision history for Microsoft Office files, stored in Google Docs, have access to offline editing with smart synchronization of offline changes, and no need to purchase Microsoft Office upgrades or deploy SharePoint®, which has added costs.

Each committee can share documents, calendars, sites, shared folders, and videos with their own committee. Administrators can control who has access to content by managing group membership and who can see what. Group discussions are archived by default, allowing users to easily search and view past and present discussions via the Web user interface (UI).

With Google Sites, you will be able to deploy an intranet and an Internet website for the organization, if you like. It will not be all that fancy, but the need for fancy is dictated by who will access it and what they want to do when they access the site. There are templates for project workspace team sites, intranets, and some just to have a simple presence on the Web.

A video utility facilitates video sharing and makes important communications such as internal trainings and corporate announcements more engaging and effective. You will be able to keep videos secure and private so that providers and administrators and support staff can securely share video trainings with co-workers without exposing confidential information.

I could go on and on about the other roles, responsibilities, and oversight you will have with this committee, but its functionality continuously evolves as the organization evolves. My intention for this chapter was to share a good launch pad for you to get you thinking about committee mission, goals, objectives, budget, capitalization, consultant selection, and timelines. I hope that I met my objective for you here. Best of luck!

Section III

Business Development

Contracting and Marketing

15 Contracting with Payer Organizations

Since early in the twentieth century, health insurance coverage has been an important issue in the United States. The first coordinated efforts to establish government health insurance were initiated at the state level between 1915 and 1920. However, these efforts came to naught. Renewed interest in government health insurance surfaced at the federal level during the 1930s, but nothing concrete resulted beyond the limited provisions in the Social Security Act that supported state activities relating to public health and healthcare services for mothers and children.

From the late 1930s on, most people desired some form of health insurance to provide protection against unpredictable and potentially catastrophic medical costs. The main issue was whether health insurance should be privately or publicly financed. Private health insurance, mostly group insurance financed through the employment relationship, ultimately prevailed for the great majority of the population.

Private health insurance coverage grew rapidly during World War II, as employee fringe benefits were expanded because the government limited direct wage increases. This trend continued after the war.

Concurrently, numerous bills incorporating proposals for national health insurance, financed by payroll taxes, were introduced in Congress during the 1940s; however, none was ever brought to a vote.

Instead, Congress acted in 1950 to improve access to medical care for needy persons who were receiving public assistance. This action permitted, for the first time, federal participation in the financing of state payments made directly to the providers of medical care for costs incurred by public assistance recipients.

Congress also perceived that aged individuals, like the needy, required improved access to medical care. Views differed, however, regarding the best method for achieving this goal. Pertinent legislative proposals in the 1950s and early 1960s reflected widely different approaches. When consensus proved elusive, Congress passed limited legislation in 1960, including legislation titled "Medical Assistance to the Aged," which provided medical assistance for aged persons who were less poor, yet still needed assistance with medical expenses.

After a lengthy national debate, Congress passed legislation in 1965 establishing the Medicare and Medicaid programs as Title XVIII and Title XIX, respectively, of the Social Security Act. Medicare was established in response to the specific medical care needs of the elderly, with coverage added in 1973 for certain disabled persons and certain persons with kidney disease. Medicaid was established in response to the widely perceived inadequacy of welfare medical care under public assistance.

Responsibility for administering the Medicare and Medicaid programs was entrusted to the Department of Health, Education, and Welfare—the forerunner of the current Department of Health and Human Services (DHHS). Until 1977, the Social Security Administration (SSA) managed the Medicare program, and the Social and Rehabilitation Service (SRS) managed the Medicaid program. The duties were then transferred from SSA and SRS to the newly formed Health Care Financing Administration (HCFA), renamed in 2001 the Centers for Medicare & Medicaid Services (CMS).

The Health Maintenance Organization Act of 1973 (Public Law 93-222), also known as the HMO Act of 1973, 42 U.S.C. §300e, is a law passed by the U.S. Congress that resulted from discussions Paul Ellwood had with what is today the DHHS. It provided grants and loans to provide, start, or expand a Health Maintenance Organization (HMO); removed certain state restrictions for federally qualified HMOs; and required employers with twenty-five or more employees to offer federally certified HMO options *if* they offered traditional health insurance to employees. It did not require employers to offer health insurance. "HMOs" were defined simply as plans that: specified list of benefits to all members, charged all members the same monthly premium, and most were structured as a nonprofit organization.

HMOs are licensed at the state level, under a license that is known as a Certificate of Authority (COA) rather than under an insurance license. This alone contributes much to the inefficiency of the healthcare reimbursement systems in the United States. UnitedHealthcare, Aetna, Cigna, and others do not have one organization through which they do business with the hospitals, and healthcare and ancillary facilities in America or the countries to which they have expanded.

As such, each state requires them to obtain a COA to permit them to engage in the business of insurance. To do this, they have to build a business corporation in each state in which they operate, follow different regulations in each state, produce separate actuarial and other reports and projections required by regulation in each state, and contract with providers separately in each state. The whole redundant system is wasteful, confusing, and expensive.

In 1972, the National Association of Insurance Commissioners adopted the HMO Model Act, which was intended to provide a model regulatory structure for states to use in authorizing the establishment of HMOs and in monitoring their operations. At its best, an HMO is a health delivery system that combines both the financing and delivery of healthcare for enrollees. Based on the form of HMO, the financing can be done through an insurance-like product that protects policy holders from the financial exposure of certain "covered expenses" deemed so under the Certificate of Coverage, which describes what the plan will pay for and under the circumstances in which payment will come from the plan, and when it will not pay.

The HMO can be structured as a network model, a staff model, an IPA (Independent Practice Association) or PHO (Physician Hospital Organization) model, or a group model.

Network Model: In this model, an HMO contracts with several groups of physicians to provide services to the HMO's members. The HMO may contract with several large multi-specialty groups or with many groups of primary care practitioners. Physicians maintain their own offices and are often compensated through a payment method known as capitation, but other reimbursement models are also available.

For a predefined amount per-member-per-month (PMPM), the group agrees to provide all needed services for a specific population of patients. Physicians in this type of arrangement are often said to be assuming "risk" because they accept the risk for the cost of care of the patients in the network. In this type of arrangement, physicians may also see patients that are non-HMO members, who are members of other HMOs, PPOs, Medicare, or workers' compensation, or self-pay.

Staff-model HMOs: Physicians are salaried employees of the HMO and are hired to care for the HMO's patients/beneficiaries; they do not see other patients outside the plan beneficiaries. In addition to salary, physicians may receive bonus and incentive payments based on performance and productivity.

Generally, staff-model HMOs tend to employ generalists and physicians who practice in many of the common subspecialties. In some cases, specialty physicians may be employed or contractually linked to the HMO. Physicians in this model care for a specific group of people usually referred to as a panel of patients. Both the Harvard Community Health Plan and the Group Health Cooperative of Puget Sound are examples of staff-model HMOs. Contrary to popular assumption, Kaiser Permanente is not a staff-model HMO, but is instead a group model, which I define next.

Physicians who practice are, more often than not, salaried employees. They receive a guaranteed salary, allied health support, and their work hours are fairly regular. A practice administrator handles the business aspects of the practice, leaving physicians free to do what they do best, take care of patients. Most in this setting who want this comfort zone are happy here.

Many staff-model HMOs provide services that individual-physician offices could not afford to maintain, such as lab and radiology facilities and other types of diagnostic equipment. Staff-model HMOs can exert a greater degree of control over healthcare delivery. As employees, and not independent contractors, physicians may be required to follow practice guidelines and clinical protocols established by their employers who manage and control utilization of health services.

As a salaried employee, physicians in a staff-model HMO limit their income potential. This is a trade-off that many physicians accept willingly, because it helps them manage the work–life balance. Unlike a private practice where physicians can increase their earnings by expanding hours or increasing patient load, physicians in staff-model HMOs receive a salary in return for all clinical services they perform on behalf of patients. Raises, bonuses, continuing education time off, course reimbursements, vacation times, and salary increases must be negotiated.

Group-model HMOs: Group-model HMOs may be classified as either "captive" or "independent." In the captive model, the HMO forms the group, usually a large multi-specialty group, to provide services to HMO members. In the independent model, the HMO contracts with an existing group to provide physician services to members. The Kaiser Foundation Health Plan is an example of a captive group. The Geisinger Clinic is an example of an independent group practice.

Both models, captive and independent, are referred to as closed-panel HMOs because physicians must be members of the group practice to participate in the HMO. In the Kaiser model, the doctors are employed by various local or regional Permanente Medical Group corporations. These, in turn, contract with the insurance product, and they maintain admitting privileges at the affiliated in-patient facilities owned by or contracted with the corporations.

Independent Practice Association (IPA) and Physician Hospital Organization (PHO) HMO Models: In this HMO model, the insurer or health insuring organization contracts with the IPA or PHO entity, or its Management Services Organization (MSO). Once the contract is executed, physicians provide services in return for a negotiated rate. Physicians work from their own offices and are not limited in terms of the patients they may see. In this model, reimbursement may involve capitation or discounted fee-for-service, or myriad other creative reimbursement arrangements.

This model helps physicians achieve some of the benefits of belonging to a large group for negotiating purposes while allowing physicians to maintain a great deal of autonomy in their individual practices. Crucial here is the adherence to legal and operational requirements to remain compliant with antitrust and anti-self-referral regulations. Depending on the legal form and operational characteristics, there may be some things that one IPA, PHO, or MSO may do with impunity while a competing organization described by the same set of three letters may not act in the same manner. This has to do with risk assumption, integration levels, and operating efficiencies.

That took us through the regulations established in 1973 with the HMO Act. In 1974, another regulation was passed, known as ERISA, the Employee Retirement Income Security Act of 1974. ERISA covers pension plans and welfare benefit plans (e.g., employment-based medical and hospitalization benefits, apprenticeship plans, and other plans described in Section 3(1) of Title I).

Initially, the IRS was the primary regulator of private pension plans and, later, healthcare benefits. The Revenue Acts of 1921 and 1926 allowed employers to deduct pension contributions from corporate income, and allowed for the income of the pension fund's portfolio to accumulate tax-free. The participant in the plan realized no income until monies were distributed to the participant, provided the plan was tax qualified. To qualify for such favorable tax treatment, the plans had to meet certain minimum employee coverage and employer contribution requirements.

The Revenue Act of 1942 provided stricter participation requirements and, for the first time, disclosure requirements.

The U.S. Department of Labor became involved in the regulation of employee benefits plans upon passage of the Welfare and Pension Plans Disclosure Act in 1959 (WPPDA). Plan sponsors (e.g., employers and labor unions) were required to file plan descriptions (Summary Plan Descriptions or SPDs) and annual financial reports (IRS Form 5500) with the government; these materials were also available to plan participants and beneficiaries. This legislation was intended to provide employees with enough information regarding plans so that they could monitor their plans to prevent mismanagement and abuse of plan funds. The WPPDA was amended in 1962, at which time the Secretary of Labor was given enforcement, interpretative, and investigatory powers over employee benefit plans to prevent mismanagement and abuse of plan funds. Compared to ERISA, the WPPDA had a very limited scope.

The goal of Title I of ERISA is to protect the interests of participants and their beneficiaries in employee benefit plans. Among other things, ERISA requires that sponsors of private employee benefit plans provide participants and beneficiaries with adequate information regarding their plans.

Typically, employers and unions establish a trust account to pay the costs of healthcare claims. They can be contributory and seek part of the money to fund the trust from the employees or union members. It is less costly in many cases to leave the money in the trust account, access discounts through a PPO (Preferred Provider Organization), and use some of the money in the trust fund to pay the discount access fees, a bit more to pay for the services of a third-party administrator (TPA); purchase case management, disease management, claims auditing services, and other plan operation necessities; and to use a bit more to purchase excess loss coverage known as reinsurance coverage, than it is to pay monthly premiums to transfer the risk of the cost of claims payment and administrative services to large HMOs or insurers.

A voluntary employees' beneficiary association under Internal Revenue Code section 501(c)(9) is an organization organized to pay life, sick, accident, and similar benefits to members or their dependents, or designated beneficiaries, if no part of the net earnings of the association inures to the benefit of any private shareholder or individual. The organization must meet the following requirements:

1. It must be a voluntary association of employees.
2. The organization must provide for payment of life, sick, accident, or other similar benefits to members or their dependents or designated beneficiaries and substantially all of its operations are for this purpose.
3. Its earnings may not inure to the benefit of any private individual or shareholder other than through the payment of benefits described in (2) above.

Membership of a section 501(c)(9) organization must consist of individuals who are employees that have an employment-related common bond. This common bond may be a common employer (or affiliated employers), coverage under one or more collective bargaining agreements, membership in a labor union, or membership in one or more locals of a national or international labor union. An organization that is part of a plan will not be exempt unless the plan meets certain nondiscrimination requirements. However, if the organization is part of a plan maintained under a collective bargaining agreement between employee representatives and employers, and such plan was the subject of good-faith bargaining between such employee representatives and employers, then the plan need not meet such nondiscrimination requirements for the organization to qualify as tax exempt. An organization will not be treated as exempt under section 501(c)(9) unless it gives timely notice to the IRS that is it applying for recognition of such status.

The popularity of these organizations is why there has been such momentum in the self-funded market while there has been shrinkage of market share in the HMO and indemnity insurer markets. Indemnity insurance compensates the beneficiaries of the policies for their actual economic losses, and entitles the insured to seek reimbursement up to the point of maximum medical improvement, limited only by the policy dollar limits set in the insurance policy. HMOs compensate the beneficiaries of the policies for their actual economic losses, up to the limiting amount of the benefit; but if the benefit is limited, such as is often the case in services such as ambulance, diagnostic imaging, physical therapy services, medical equipment, etc., then even if there are dollars left in the policy limit, the maximum coverage entitlement is the benefit limit, rather than the policy limit, even if maximum medical improvement has not yet been reached.

Individuals who manage plans (and other fiduciaries) must meet certain standards of conduct, derived from the common law of trusts and made applicable (with certain modifications) to all fiduciaries. The law also contains detailed provisions for reporting to the government and disclosure to participants. Furthermore, there are civil enforcement provisions aimed at ensuring that plan funds are protected and that participants who qualify receive their benefits.

Plan sponsors must design and administer their plans in accordance with ERISA. Title II of ERISA contains standards that must be met by employee pension benefit plans in order to qualify for favorable tax treatment. Noncompliance with these tax qualification requirements of ERISA may result in disqualification of a plan and/or other penalties.

Employee Benefits Security Administration (EBSA) responsibilities under ERISA were expanded by healthcare reform. The Consolidated Omnibus Budget Reconciliation Act of 1985 (COBRA) added a new Part 6 to Title I of ERISA that provides for the continuation of healthcare coverage for employees and their beneficiaries (for a limited period of time) if certain events would otherwise result in a reduction in benefits.

More recently, the Health Insurance Portability and Accountability Act of 1996 (HIPAA) added a new Part 7 to Title I of ERISA aimed at making healthcare coverage more portable and secure for employees, and gave the department broad additional responsibilities with respect to private health plans.

In the strictest of managed-care models, each member is assigned a "gatekeeper," usually a primary care physician (PCP) who is financially responsible for the healthcare expense management and cost containment for the overall care of members assigned to him or her. Specialty services require a specific referral from the PCP to the specialist. Non-emergency hospital admissions also required specific preauthorization by the PCP. Typically, services are not covered if performed by a provider not an employee of or specifically approved by the HMO, unless it is an emergency situation as defined by the HMO. Financial sanctions for use of emergency facilities in non-emergency situations were once an issue; however, prudent layperson language now applies to all emergency-service utilization and penalties are rare.

Before managed care, indemnity insurance paid any hospital or physician chosen by a patient on a cost-plus or fee-for-service basis. Managed care brought about a change in contracting and reimbursement for hospital services. Unlike indemnity insurers, managed care plans selectively contract with hospitals in order to negotiate lower hospital prices, shift payment risk to hospitals, and form provider networks that appeal to their enrollees. Managed-care plans extract price discounts by threatening to exclude providers from their selected networks. Hospitals and physicians wanting to improve their competitive position for managed-care contracts try to lower costs or develop strategies to counter managed-care bargaining power.

One response to the rise of managed care was for hospitals and physicians to integrate into IPAs, PHOs and MSOs, and Clinics without Walls. I personally managed more than 150 projects in the United States and have since managed the design, development, launch, and shepherded the first-year activities on a consulting basis of several of these abroad. Others on our team of specialized managed-care consultants with the Mercury Healthcare Advisory Group have assisted others in the design, operation, and even preparation for accreditation by international and U.S. quality and safety accreditation schemes such as the Joint Commission and the Joint Commission International, and many others.

One response to the rise of managed care is for hospitals and physicians to integrate. The literature provides two explanations for why hospitals and physicians have formed vertical relationships in response to managed care. The first is a transactions costs argument that such relationships increase efficiency and quality. With greater efficiency, providers are able to offer managed-care plans at lower prices without sacrificing quality. The second is that hospitals and physicians ally in order to improve their bargaining position with managed-care plans and other insurers, and thereby raise or stabilize prices.

Before managed care, under fee-for-service payment to physicians and cost-plus payment to hospitals, there was little financial incentive for hospitals and physicians to work together to achieve economies of scope and otherwise become more efficient. In a managed-care environment, where providers are paid via capitation and other forms of prospective payment,* physicians and hospitals can accrue the financial benefits of increased efficiency if they can overcome internal agency problems and take advantage of economies of scope.

Vertical integration may be better able to take advantage of possible economies of scope. The care of any one patient typically spans both hospital and physician office settings in the regional or local market, and can span countries in a health travel or expatriate market. By changing the process of patient care and coordinating care across sites, joint hospital-physician organizations may be more efficient. Shared information systems can be put into place to gather data on costs, quantity, and quality, and monitor performance relative to benchmarks. Integrated management can facilitate the sharing and use of information and identify areas of complementarities and substitutability. It is this premise that will withstand the test of time, and it is this premise upon which the basis of the globally integrated health delivery system® is designed. Regardless of what happens in the political arena, these basic core functions are needed to care for patients and coordinate health logistics for a mobile workforce that includes traveling employees, expatriates, domestic workers, and retirees with extended healthcare benefits paid for by corporations. The rest is simply geography.

Since the 1980s, under the ERISA Act passed in Congress in 1974 and its preemptive effect on state common law tort lawsuits that "relate to" Employee Benefit Plans, HMOs administering benefits through private employer health plans have been protected by federal law from malpractice litigation on the grounds that the decisions regarding patient care are administrative rather than medical in nature.†

While managed-care techniques were pioneered by HMOs, they are now used by a variety of private health benefit programs. Managed care is now nearly ubiquitous in the United States; but if one travels as I have, one finds that managed care

* Other forms of prospective payment include paying a fixed fee to treat an illness, such as Medicare Diagnostic Related Groups or a fixed per-diem. In these cases, the provider is at some financial risk for the costs of care.

† See *Cigna v. Calad*, 2004.

is thriving and flourishing in other countries as well. In the United States, managed care has faced backlash because it has largely failed in the overall goal of controlling medical costs. I maintain that a large part of the contributory liability for the failure is from the sacred state-by-state silos engineered by the HMO Act of 1973, and the state insurance and HMO acts each requiring billions in administrative redundancy and fragmentation that impedes continuity of care, claims administration efficiency, and expense that does nothing to effectively contain costs, while preserving profit margins and fat salaries and benefits to health plan executives, regardless of for-profit or not-for-profit status.

Just as the PPACA solidified the term "ACO," the HMO Act solidified the term "HMO" and gave HMOs greater access to the employer-based market. ERISA has reset this advantage to the employer, who now far outnumbers the HMOs and insurance companies operating in the group health market.

Currently, in the United State, more than 210,000 of these employer plans exist. Another several thousand are sponsored by union health and welfare trusts. They are free to contract directly with hospitals, IPAs, PHOs, MSOs, and other provider organizations or to contract with TPAs who might negotiate discounts on their behalf, or with PPOs who contract with providers, create a network of providers, and then sell access to the discounts available through the network on a per-employee-per-month (PEPM) basis.

In comparison, in 2011, the National Committee for Quality Assurance (NCQA) ranked 830 health plans in all 50 states plus the District of Columbia. There were 390 plans for people who obtain insurance through their employer, 341 for people with Medicare, and 99 for those covered by Medicaid.* IPAs, PHOs, and MSOs are eligible to contract with these plans too if they can attract the interest of the payor organization with their USP. For a globally integrated health delivery system®, the market is less likely to be these redundant silos of insurance and HMO plans, and more likely to be the self-funded ERISA multinational or global employer that has workforce spread globally, and must manage costs, logistics, and healthcare continuity of care in places that may be remote, have inadequate care, or limit or restrict expatriate freedom of medical care by certain cultural and other laws, especially those that relate to women or the LGBT† workforces.

Health spending in the United States has reportedly grown rapidly over the past few decades. From $27.5 billion in 1960, it grew to $912.5 billion in 1993, increasing at an average rate of 11.2 percent annually. This strong growth boosted healthcare's role in the overall economy, with health expenditures rising from 5.2 percent to 13.7 percent of the Gross Domestic Product (GDP) between 1960 and 1993. According to many sources and a search using Google, these numbers are readily available and cited by numerous sources. It is difficult to pinpoint to whom credit should be given for the citation, but at this point the numbers are in the public domain and attributed to numerous government sources, including the Centers for Medicare and Medicaid Services, the Office of Management and Budget, the

---

* http://www.consumerreports.org/health/insurance/health-insurance/health-insurance-rankings-data/index.htm [doi: 1/8/2012].

† Lesbian, Gay, Bisexual, or Transgendered.

Department of Health and Human Services, and hundreds of others. To cite them all here would be pointless.

Between 1993 and 1999, strong growth trends in healthcare spending subsided. Strange … that was about the time we were involved in healthcare reform during the Clinton Administration. Over this period, health spending rose at a 5.6-percent average annual rate to reach nearly $1.3 trillion in 1999, and the share of GDP going to healthcare stabilized, with the 1999 share measured at 13.5 percent. This stabilization reflected the nexus of several factors: the movement of most workers insured for healthcare through employer-sponsored plans to lower-cost managed care; low general and medical-specific inflation; excess capacity among some health service providers, which boosted competition and drove down prices; and GDP growth that matched slow health spending growth.

Between 1999 and 2002, growth picked up, averaging 8.2 percent annually. During this period, the share of GDP devoted to healthcare increased from 13.5 percent to 15.1 percent. Health spending grew more slowly after 2002, averaging 6.2 percent annually from 2003 to 2008, as its share of GDP increased from 15.6 percent to 16.2 percent. In 2008, health spending reached $2.3 trillion, or $7,681 per person. This is a key figure because it relates to the new limit under the Patient Protection and Affordable Care Act (PPACA) reform's so-called "Cadillac plan" excise tax, which is likely to affect more than 60 percent of large employers' active health plans by the provision's 2018 effective date.

The excise tax was included in the Patient Protection and the PPACA passed into law on March 23, 2010. The provision levies a 40-percent nondeductible tax on the annual value of health plan costs for employees that exceed $10,200 for single coverage or $27,500 for family coverage in 2018. The average 2010 cost of medical coverage for active single and family plans is $5,184 and $14,988, respectively. When these figures are projected out to 2018 with reasonable estimates of future healthcare inflation, the excise tax is often triggered.

As a result of the excise tax provision, a plan with single coverage costs of $11,200 in 2018 would exceed the limit by $1,000 and be assessed a tax of $400. If 10,000 employees were enrolled in that plan, the total tax bill would be $4,000,000. The tax is paid by the employer, either through increased premiums on an insured plan or a surcharge levied by the administrator of a self-funded health plan. Employers will be forced to either absorb the additional tax or pass some, or all, of it back to employees in the form of higher premiums. What it would not do is purchase additional healthcare or benefit the plan participants in any way.

A classmate (whose name I did not catch, but who works for Towers Watson as an actuary) in my insurance producer license training class in December 2011 stated that all it would take to drive costs above the excise tax cap for six in ten employers is an 8-percent average annual cost increase. Without making plan design changes, that is what many employers are projecting. They need a solution, and the current Medicare and Medicaid reforms do not seem to offer a complete fix for their expatriate workforces, and do not really involve them in ACOs (Accountable Care Organizations), so they really need another market-driven solution—one that reduces the privately funded healthcare, including employer cost shares, individuals' out-of-pocket expenditures, private health insurance supplements such as international private medical insurance purchased either by or for their expatriates, and the cost of travel accident and illness insurances and workers compensation coverage for their employees, dependents, and retirees.

Still another source of payment for healthcare comes from philanthropy and non-patient revenues (such as revenue from gift shops and parking lots), as well as health services that are provided at no cost through charity programs. Another market segment that we do not often hear about in all the talk of healthcare reform is that of inbound health travel from governments and wealthy individuals seeking care in the United States from other countries where healthcare may be scarce, unavailable, inadequate, or difficult to access because of long waits or services not permitted by law.

A McKinsey and Co. report from 2008 found that a plurality of an estimated 60,000 to 85,000 medical tourists were traveling to the United States for the purpose of receiving in-patient medical care.[*] The availability of advanced medical technology and sophisticated training of physicians are cited as driving motivators for growth in foreigners traveling to the United States for medical care.[†]

Public spending represents expenditures by federal, state, and local governments. A significant portion of public health spending can be attributed to the programs administered by the Centers for Medicare & Medicaid Services (CMS)—Medicare, Medicaid, and the Children's Health Insurance Program (CHIP, known from its inception until March 2009 as the State Children's Health Insurance Program, or SCHIP). Together, Medicare, Medicaid, and CHIP financed $823.8 billion in healthcare services in 2008—slightly more than one-third of the country's total healthcare expenditures and almost three-fourths of all public spending on healthcare. Since their enactment, both Medicare and Medicaid have been subject to numerous legislative and administrative changes designed to make improvements in the provision of healthcare services to our nation's aged, disabled, and disadvantaged. A significant example is the Medicare Prescription Drug, Improvement, and Modernization Act (MMA) of 2003 (Public Law 108-173), which created the Medicare Advantage program and provided Part D prescription drug coverage for Medicare beneficiaries beginning in 2006.

This Act not only established the Medicare Advantage program, but also established various Consumer-Driven Health Care programs such as Health Savings Accounts (HSAs), Healthcare Reimbursement Arrangements (HRAs), and Special-Purpose HRAs. Employers and small businesses alike can use these programs to fund discretionary health expenses for some plan participants without having to calculate the potential claims risk exposure for a limited-use benefit that might be used by only a few members of the plan, for example, assisted reproductive therapies, maternity surrogacy, health travel options, and other costs associated with wellness initiatives. Because those are covered in my managed care contracting handbook[‡], I will not duplicate those extended explanations in this book.

---

[*] U.S. Hospitals Worth the Trip, Forbes, http://www.forbes.com/2008/05/25/health-hospitals-care-forbeslife-cx_avd_outsourcing08_0529healthoutsourcing.html [doi, January 8, 2012].

[†] Ibid.

[‡] Todd, Maria K., *The Managed Care Contracting Handbook, 2nd edition*, Planning and Negotiating the Managed Care Relationship (2011, Productivity Press, New York), pp. 55–62.

The remaining portion of publicly funded healthcare spending in the United States amounted to 823.8 billion in 2008 and includes expenditures for the following: the Department of Defense healthcare program for military personnel, the Department of Veterans' Affairs health program, noncommercial medical research, payments for healthcare under Workers' Compensation programs, health programs under state-only general assistance programs, the construction of public medical facilities and the purchase of equipment, maternal and child health services, school health programs, subsidies for public hospitals and clinics, Indian healthcare services, substance abuse and mental health activities, and medically related vocational rehabilitation services.

16 Contracting for Capitation and Bundled Service Arrangements

In most managed-care organizations, covered medical and related services are usually rendered and payments are based upon a prospective budget allocated as a percentage of the premium collected by the health plan and are often referred to as a "target medical cost index" on a per-member, per-month basis. Usually, actuaries assist both managed-care payors and providers in the measurement of risk exposure by statistically analyzing historical utilization data supplied by organizations such as the Health Insurance Association of America (HIAA). The data are derived by encounters reported via claims for payment of medical and related services. A problem with this system of measurement is that the data are somewhat skewed for at least three reasons:

1. *Data extrapolation.* Data are prepared into reports by various software programs that organize the data into nonstandard formats.
2. *Data entry errors.* Errors occur either in claims processing or claims submission.
3. *Fraud and abuse.* Coding aberrations are used to manipulate payment for claims that might otherwise go unpaid or be paid at substantially lower rates.

CAPITATION

In highly managed plans, revenue is allocated using a prospective method of payment known as *capitation.* Capitation is a system of reimbursement that provides financial incentives and disincentives related to the use of specific providers, services, or service sites. In capitated managed-care plans, a contract is negotiated for a specific menu of services, and a fixed amount of money is paid to the provider of care in anticipation of rendering those services to patients who have selected that provider. The money is paid in advance of the need for the services on a per-member-per-month (PMPM) basis and usually only varies by the age and the sex of the patient. It is designed to be independent of the actual volume or cost of the services rendered. In this way, the provider assumes all risk for high incidence and inefficient/ineffective medical management (see Figure 16.1).

Capitation Demographics Analysis

The capitation calculation takes the following into account:

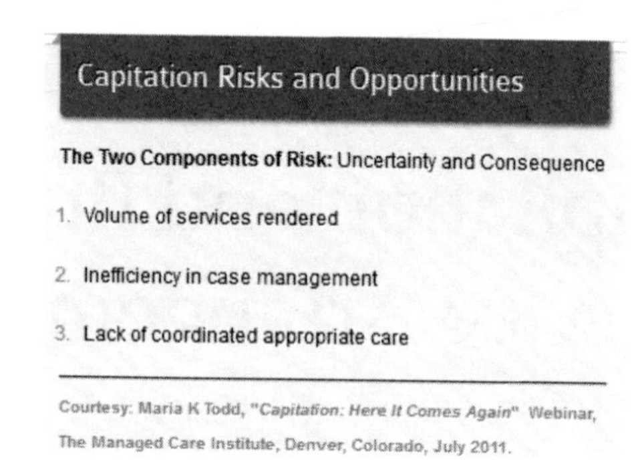

FIGURE 16.1 Risks and opportunities of capitation.

- Morbidity
- Mortality
- Lifestyle
- Age
- Gender
- Education
- Occupation
- Socioeconomic status
- Historical utilization of services

In previous years, we studied capitation and came to a comfortable point in predictability of risk using a critical mass of patients from a typical case mix that included both users of services and non-users. More recently, however, with the advent of Consumer-Driven Health Plans (CDHPs), the healthier populations are now joining the CDHP programs and are no longer mixed as they were in years past in the HMO (Health Maintenance Organization) managed-care risk pools.

This leaves a strange new situation that has not been addressed much in the current literature. We used to have the consumption of services offset by those on the roster that did not consumer services, by paying into the premium pool but not incurring claims. Therefore, they offset the use of claims dollars in the risk pools for those who for one reason or another spent more than their share of the dollars.

In current times, those non-users have been moved to a different funding mechanism and the HMO managed-care population contains more users than in years past. This will undoubtedly change the dynamics of the critical mass needed in order to manage capitation and claims risk to the degree that we were able to in years past. We will have to look to our learned actuaries for guidance and lean on excess loss coverage like never before to bolster us financially as we navigate the learning curve.

Services

The capacity and capability of the providers also impact the capitation. Can the hospital/provider provide all services, or must some of the services be "carved out" or subcontracted? What services beyond office visits are included: lab, physical therapy, in-office surgery, radiology, injections? What are the referral arrangements with subspecialists? What is the mix of family practitioners to general internists?

DEALING WITH UNPREDICTABLE AND UNMANAGEABLE RISK REINSURANCE

In managed care, reinsurance is a type of insurance that holders of risk purchase to guard against unpredictable risk or unmanageable risk and/or low enrollment. Industry experts used to concur that groups that accept capitated risks should purchase their own reinsurance when their rosters reach more than 2,000 covered lives. The sentiment has not changed, but the minimum critical mass has, and no one has been forthcoming about what the new number is. This author maintains that the reason for this is because of the shift in risk associated with the types of lives that remain in the risk pools now that the healthier populations have moved to HRAs (Healthcare Reimbursement Arrangements) and HSAs (Health Savings Accounts) not always associated with HMO plans.

One of the ways providers can protect themselves against excess utilization costs is to purchase reinsurance or stop-loss coverage. The reinsurance carrier generally prices its coverage at twice what it expects to pay out over a policy period.

Hospitals generally choose to purchase reinsurance using per-diems or some other factors as accumulators. A $25,000 deductible may cost an average of 7 percent of the hospital capitation. Other providers may elect to cover catastrophic incidences, using charges as accumulators with a deductible of $150,000, paying about 1 percent of the capitation. Now that the minimum critical mass has changed, all those numbers are meaningless. I have not been able to ascertain from any resources what the new numbers are or will be in the short term.

Reinsurance should be used when the provider has low enrollment, is responsible for costs over which it has little or no control, or is dealing in a large number of unknowns. Because we no longer have an industry standard for what "low" enrollment is, it is difficult to provide resources or guidance in trade books such as this. My best advice is to find an actuary who you trust and a reinsurance broker who will work with you, demonstrate patience, and guide you through the process using relationship as the source of value instead of just a quick commission for a policy sale. There are many brokers out there who are just waiting to assist their clients through this learning curve. Shop carefully and diligently!

DIVIDING THE PIE

Managed-care plans take into account many snippets of data in order to contemplate and create an offer of a contract for capitated reimbursement to providers. One thing they take into account prior to extending an offer is how not to pay for upside risk, and how not to get caught in too much downside risk, such that there is nothing to return back to Wall Street.

Many plans pay providers on the basis of a percentage of premiums, based on historical expenditures tempered by where they are in the underwriting cycle. The typical arrangement is illustrated in Table 16.1.

TABLE16.1
Capitation Development Analysis

Service	Annual Freq./1,000	Avg. Charge	Capitation PMPM
	Primary Care		
Ambulatory office visits	2.55	$41.00	$8.71
Inpatient visits	0.116	$62.00	$0.60
Home visits	0.005	$65.00	$0.03
Pathology	1.536	$10.00	$1.28
	Misc. Office Services/Immunizations		
Injects, incl. allergy	0.6	$18.00	$0.90
EKG, audiometry, etc.	0.25	$38.00	$0.79
Office surgery	0.1	$75.00	$0.63
Total Primary Care			**$12.94**
	Referral Care/Services		
Ambulatory office visits	0.85	$48.00	$3.40
Inpatient visits	0.115	$75.00	$0.72
Home visits	0.003	$95.00	$0.02
Consultations	0.05	$135.00	$0.56
Misc. office services	0.5	$50.00	$2.08
Emergency room visits	0.15	$80.00	$1.00
	Surgery		
Inpatient	0.068	$675.00	$3.83
One-day surgery	0.045	$425.00	$1.59
Office surgery	0.15	$110.00	$1.38
Anesthesia	0.072	$372.00	$2.23
Maternity	0.072	$1,640.00	$3.01
Radiology	0.58	$124.00	$5.99
Outpatient psychiatric	0.2	$68.00	$1.13
	Hospital Outpatient Svcs.		
Physical & radiation therapies	0.06	$90.00	$0.45
Ambulance	0.01	$195.00	$0.16
Home health care	0.02	$58.00	$0.10
Durable medical equipment			$0.45
Prescription drugs	4.5	$27.00	$10.13
Total referral care	$38.23		
	Hospital Services		
Inpatient care	0.35	$1,100.00	$32.08
One-day surgery	0.045	$1,250.00	$4.69
Extended care facility	0.015	$ 225.00	$0.28
Emergency room	0.18	$190.00	$2.85

TABLE16.1 (*Continued*)
Capitation Development Analysis

Service	Annual Freq/1000	Avg Charge	Capitation PMPM
Total Hospital Care			**$39.90**
Total Medical Capitation			**$91.07**
Reinsurance/Catastrophic Allowance		**$ 4.93/PMPM**	
Plan Margin 20%		**$24.00/PMPM**	
Total			**$120.00/PMPM**

Source: Courtesy Maria K Todd, Mercury Healthcare Advisory Group, Inc.

How to use this table:
(All numbers are fictitious; they are simply placeholders to walk you through the arithmetic modeling.)

- Assume that primary care ambulatory visits = 2.550 visits per year.
- Assume that the average cost to HMO per visit = $41.00 per visit.
- Calculation: 2.550 × $41.00 = $104.55, divided by 12 = $8.71

The utilization numbers must be provided by an actuary by having the information required at the beginning of the capitation section of this chapter. Next, the numbers in the middle column need to either be based on the provider's target reimbursement or the plan's reimbursement as set forth in their fee-for-service equivalent.

In essence, multiply the real number you are provided or can historically derive on the left column, by the number you establish as correct in the middle column, divide by 12 and you should come up with the needed capitation to meet the projected utilization to cover prepayment of those services at the target reimbursement rate. If either number is off, you have risk or you will find yourself "in the money." More frequently the former than the latter.

The significance of correct age, enrollment size, health status, covered services, and gender weighting of capitation cannot be overemphasized. Without these specifics, readers are cautioned not to use Table16.1 to negotiate capitation rates.

Once a premium has been established, a capitation development analysis is prepared by the actuaries. A typical commercial capitation analysis looks like the table above. Through the use of this table, payers and providers alike may forecast utilization and payer cost indices under their fee allowances to estimate capitation equivalencies.

FEE SCHEDULES

For covered medically necessary services that are not included in the capitation budget, a fee schedule may be utilized to determine reimbursement. Fee schedules are usually a product of the multiplication of one of a variety of reference systems of relative value units by a conversion factor represented by dollars, which, when multiplied together, determine the price to be paid for a specific service.

Fee schedules are sometimes manipulated by payors using "cost-fixed" or self-authored codes to represent hybridized bundled services that have no value units established. These are troublesome to coders and actuaries because of their nonconformity to traditional calculations in comparison with established procedure code nomenclature relative value units.

Many hospitals are asked to consider reimbursement based on Diagnostic Related Groups (DRGs). At the time of this manuscript, the final Inpatient Prospective Payment System reimbursement has just been released. A cursory review led me to some immediate key observations for hospitals that currently have contracts with Evergreen clauses ("rollover features") that do not escalate rates or keep them contemporary with current regulatory recalculations, and this one will be significant.

Providers should revisit existing contracts as well as remodel new proposals with the following question in mind:

> You will need to consider the continued transition from charge-based to cost-based reimbursement: How will this translate into modeled reality for your existing contracts that are tied to Medicare rates prior to the new DRG system?

I have not seen current proposals and rate sheets that permit many fees to be modeled to address the new methodologies for calculating outlier payments and capital cost reimbursement. Be sure to address those in detail with your health plan contracting representatives, and obtain all explanations and model arithmetic in writing, which can be attached to the contract to prevent an argument of hearsay in accordance with the Entire Agreement provision of your contract.

Treatment of recalled medical services and devices, as well as replacement surgeries/procedures were not addressed at all in any of the recent contracts reviewed by me.

CASE RATES

A variation on the theme of DRGs is case rate reimbursement. Based on my experience in reviewing contracts for many years, I have come to the conclusion that most providers in the United States have no clue how to negotiate a proper case rate with a health plan or any other payor, and the odds are most often loaded in favor of the plan, not the physicians or hospital.

What defies all logic to me is how the state Departments of Insurance permit these reimbursement arrangements without requiring that the providers obtain a license to be in the business of insurance without the necessary data, reinsurance or excess loss coverage, and cash reserves. My guess is that they simply are unaware of the existence of the deals because they are in service to premium payers, and the premium payers are not complaining if the hospitals and providers lose money or are underpaid. They will if the providers close up shop and close the doors!

To begin, case rates generally are designed to provide reimbursement for an "episode of care." The very term "episode" means an occurrence, incident, or event. Seems simple enough …. So when one is asked to reimburse a provider to cover an episode of care, when is the specific moment that the episode begins and ends? If it is ill-defined, or ambiguous, it could be point in time where care is delivered.

I can think of several examples that I witness in contract proposals time and time again, including

- Maternity cases,
- Outpatient surgical cases, and
- Rehabilitation cases,

to name just a few.

What is worse is that the entire concept of pricing transparency surrounding CDHP expenditures and personal accountability for healthcare expenditures is woven around online arguments of case pricing. I have a real problem with published pricing and gold stars that are "irresponsibly" posted on a website by a health plan or its agent for the purpose of steering consumers to one provider over another. Many of the postings are inaccurate, out of date, and for the most part, subjective and not exact comparisons of case rates from provider to provider as a standard.

Most contracts provide for no first right of review or the right to refuse to be marketed inaccurately, or to besmirch one provider in favor of another on the basis of price. Furthermore, most of the websites, industry articles, and speakers coming from industry tend to use a misnomer of "cost comparison" instead of a more accurate term of "price comparison." Until we can know our costs in this industry, which is wishful at best, it is impossible to label a case cost, let alone explain that cost to consumers so that they can understand what it takes to keep a medical group's doors or a hospital open for business. But alas, I digress. Let's get back on target. (Yes, I feel better now!)

So, we were in the beginning of the discussion on case rates and episodes of care as they relate to contracted reimbursement methods. In order to establish a case rate, we first need to define an episode of care, where it begins, and where it ends. Some cases should never begin, and some cases never seem to end with managed care. Take, for example, an outpatient surgery set for case rate reimbursement where the patient is a hemophiliac and will require Factor K, a very high-cost biological. If no carve-out is stated for the Factor K, the case rate will likely be in the red before it has been scheduled by the surgical scheduler.

As a best practice, both hospital and physician should take a moment together to jointly determine which cases they will entertain as a case rate and then, case by case, determine when the case rate should not be applied in favor of an alternate reimbursement methodology. Start with just five cases. Determine the case by the CPT code, taking into account instrumentation, supplies, surgical approach, OR (operating room) and anesthesia time, recovery time, and potential for adverse events during surgery and afterward.

As a former OR tech, we can take five categorical cases in outpatient orthopedic surgery. "Categorical" because that is how I repeatedly see them presented in contracts for case rates. At worst, they are categorically described as "outpatient surgery – orthopedics," "outpatient knee surgery," etc. At best, they are often only described by Ambulatory Surgical Center (ASC) Groupers (there are nine) and then arbitrarily parked into some reimbursement schema.

Let's examine these five to begin:

1. Open Reduction, Internal Fixation of a Fracture (ORIF)
2. Arthroscopic knee surgery – Meniscus repair

3. Arthroscopic knee surgery – ACL reconstruction
4. Arthroscopic shoulder surgery – Bankart procedure
5. Removal of hardware

So, you have contracted for case rates for $1,800 for each of these cases because you feel that there is some margin in doing them at that price, or because that was what the payer offered in its contract rate proposal as a percentage of Medicare rates, and you had no other interest—in time, software, or skill—to model it differently.

The typical ORIF takes about ninety minutes of OR time with a slow surgeon. The typical knee surgery takes ninety minutes, shoulder takes ninety minutes, and removal of hardware takes forty-five to sixty minutes. As far as equipment and supplies, the first four usually open up a major ortho pack, two gowns, three pairs of gloves, and assorted sutures, tapes, and dressings. The removal of hardware is generally a minor pack at best.

Anesthesia is usually MAAC or local or nerve block, and sometimes general anesthesia. If Meperidine HCL and Promethazine are used as anesthetics, there is considerable barf time and increased PACU nursing care budgeted in recovery as opposed to a quicker recovery with Propofol® (but in patients allergic to eggs, we do what we have to). I find it amusing that surgeons who own their own surgical suites rarely know the cost of each minute of OR and PACU time, and think they are saving money using less-expensive drugs for anesthesia instead of maximizing throughput of the space and staffing resources they have with quicker, more costly drugs that recover the patient faster and more compassionately—but that's part of case cost too.

One other element that is part of the case cost is the instrumentation selection and all other supplies that are opened and tossed (they essentially are "pitched" onto the back table) onto the surgical field, back table, and Mayo stand. These instruments must be counted, sterilized, unpacked prior to the case, tossed if consumable, and resterilized, packed, and prepared for the next case. Therefore, in order to do multiple cases strung together one after another, you either have to have duplicate sets of instruments, or schedule the cases accordingly so that you have time to process the equipment between cases. That all has a cost implication … and a quality implication too! (longer anesthesia time and extended surgery time, etc.)

My experience in surgery tells me that many surgical preference cards for the surgeons are likely out of date and have not been recently updated to reflect what they really want or need for the case. Therefore, many things will be tossed onto the surgical field that take time, effort, and staff to count. Some things that have been supplanted by newer things still remain, the Charge Description Master (CDM) has neither been completely updated to reflect pricing for the new items, nor to retire the no-longer-used items. It probably also means that the costs attributed to the typical case have not been remodeled recently and until these things are updated, the case cost remodeling will not occur.

All this is fine and dandy, but it is still based on that hypothetical "typical" case. Now comes the aberrance. A hemophiliac male patient is scheduled for surgery. Much more case management is needed; biological supplies such as expensive Factor K will be required. Pre- and post-operatively, he will require additional care and monitoring to ensure hemostasis. He may end up in ICU. Let's take it a step further.

Let's assume it was the ORIF case, and in addition to the surgery and the comorbid condition, he experiences a non-union of the fracture requiring additional returns to the OR for additional surgeries. In one of these, he becomes septic and infected—a true Murphy's Law case.

Issue 1: *Should this case have even been considered a case rate case?* Or should the contracted have excepted it by example, condition, or some other indicia to trigger a different reimbursement methodology? Should any of these other subsequent triggering events or that of a malignant hyperthermia or other condition that might be encountered during surgery have given rise to an abandonment of the case rate methodology and a move to a different methodology?

Issue 2: *How detailed does the contract read?* Does it even address this potential situation? Who gets to decide the final financial outcome? Based on what data or standards? Where are they published? Can they be unilaterally changed? Is the contract "mute" to this? Is the answer on the website in some obscure documents that may have never been reviewed? Have they been virtually referenced by the plan in the contract, but nobody ever checked to see if the documents really existed and what was contained in them? Assuming they were published, can the documents' established policies and coverage limitations be unilaterally amended at the sole and absolute discretion of the plan without any notice of the amendment to the provider? And if you took the case to dispute resolution, the arbiter can only decide the case based on what the Agreement states to begin with. So if you spend the money to file the dispute but you cannot prove that something is wrong, relying solely on what is stated in the contract, can you win? (You probably don't have a snowball's chance ….)

Issue 3: *How will the payor treat the treatment failure of the post-op infection?* And the cost of the associated polypharmacy and likely intravenous therapy will be required to treat it. Who pays that bill? How will you score on quality points, and will this injure your report with the little gold stars on the plan's website? What about any metrics being followed for pay-for-performance bonuses? Will this injure your position? How will this messy outcome be addressed on a weighted scale or balanced scorecard? Who has the discretion to score the outcome? Can you appeal the scoring methodology and statistical relevance?

Issue 4: *Price transparency.* You quoted this hemophiliac a price to come to your hospital. He planned to use his limited, banked, underfunded HSA (Health Savings Account) and employer-contributed HRA (Healthcare Reimbursement Arrangement) funds to cover the front-end loaded, high deductible, which at your hospital and by your staff physician was projected to come out less costly to him with a seemingly lower projected total price than the competitor up the street that was also "in network." After all, he was not planning on breaking his femur! Because he is not buying a car here, there is no list of options on the semi-private or private room window that faces the nurses' station from which to choose if he wants it or not. The fact is that he signed the consent form and stuff happens! Now there is a significant difference in price.

Oh, but wait! Although his funding mechanism is a CDHP, it is an "all products" contract so he is only facing the maximum payout that the contract states is payable for the HMO and PPO folks. So who cares? His bill is $1,800; it says so in the contract! Cool! And guess what? He has $1,700 of it in the HSA fund. But now he is angry. There was a less-than-stellar outcome. He has missed work and he has had doctor visits, and he hurts, and the law says he does not have to choose to use his HSA fund money to pay YOU! He can keep it, bank it, roll it over to next year, or just choose to be a slow pay. What is even worse is that your contract forbids you from asking for estimated patient financial responsibility up-front so not only do you have this runaway case expense, but you now have no money. Nada, zip, zero! The plan will not pay you because the amount is within the high-deductible range, so they do not have to!

The Moral of the Story

Case rate contracts are extremely complex and should never be entered into without firm modeling, strategy, a plan for implementation, and specific limitations and exclusions, in addition to well-defined beginnings and endings of an episode of care. In addition, you need to determine so many of the things I brought up in the issues above, including payment and coverage policies, adverse outcome review and appeal rights, quality grades and publicity, pay-for-performance scoring methods and bonus calculations, and separate CDHP contracts. Oh, and then there is the clinical aspect of case rates: Are the surgeon's preference cards current? Is the instrumentation reduced to only what is necessary? Have you eliminated all waste from the case? Is the CDM current? Is the case modeled appropriately for costs, and is there any other process improvement potential? I am asked to consult with hospitals all the time in this strategic development of case costs. What is amazing to me is that those who ask for the assistance to structure such a program are far outnumbered by others who have signed these case rate agreements without any more preparation than accepting Medicare ASC Rates and moving to the next line item. If you are in the latter group, I implore you to at least examine a world where the homework is done.

Now let's chat about maternity rates. Most maternity case rates are based on DRG 374, normal newborn. First, I find it humorous that the average Medicare beneficiary is over 65, so that a Medicare DRG for any infant born to a mother over the age of 65 could be classified as "normal." To quote Joan Rivers, "Can we *tawk*?" In the last thirty contracts I reviewed for clients, not once has there been mention of a different rate for vaginal births in comparison to Cesarean births. One involves lots of pushing, breathing heavy, panting, sweating, and cursing, and threats of never … well, you get the picture. The other typically involves an epidural injection, knives, assistants at surgery, retractors, scissors, tapes, sponges, an OR, an OB pack, instrument tray, and nursery space for one normal, healthy baby.

Why is it that more than 95 percent of the contract proposals for maternal-child case rates that I reviewed in my entire career as a contract analyst and managed-care expert never anticipate or mention rates for multiple births, stillborn babies, cases where the mother dies, or cases where the baby is tied to the father's plan as primary in accordance with the "birthday rule"? Is it that the plan they forgot? Or is it that they never intended to pay for more than one normal newborn baby, regardless of the outcome of the case or the number of babies delivered? And why do hospitals allow those contracts to be signed as proposed?

Bundled Case Rate / Facility & Professional Fees

CPT CODE:	29877	
Description:	Knee Surgery-Arthroscopy	
Includes:	Surgeon's History and Physical	
	Surgery	
	90 days follow-up care - limited to evaluation and management services	
	Surgical assistant, if requested	
	Anesthesiologist	
	Intraoperative x-ray and fluoroscopy, if required	
	Intraoperative gross and anatomical pathologist review, as required	
	Pre-operative testing limited to:	
		finger-stick glucometry, HCG pregnancy test, if indicated, Spin Hematocrit, Urinalysis, EKG.
		NOTE: All other lab testing and pathology at XYZ Standard Fee Schedule.
	Facility charges for routine procedure	
	Recovery Room	
	Repricing from individual bills to case rate	
	Disbursement to individual providers and MSO	
Excludes:	*Any hardware or implants	Fee: Invoice cost + 20% handling fee (however, this may be expressed in a myriad of ways that does not require the provider to tender an invoice.)
	*Facility overnight	Fee: $XXX
	*Any lab tests or follow-up x-ray or pathology services not included above	
	*Emergency transfer fees by ambulance, as required	
	*Physical therapy and DME fees, if any	
	*Transition to inpatient status for any reason shall exclude this case from eligibility under this case rate program and revert the case back to traditional workers' comp scheduled rates and reimbursement conditions.	
Price:	$xxxx	

Courtesy: Maria K Todd, Mercury Healthcare Advisory Group, Inc.

FIGURE 16.2 Rates and fees.

In conclusion to my discussion of case rates, I have included as an example cited in Figure 16.2, an example of how I prepare the layout for case rate contract negotiations by procedure. It is not exhaustive by any means, but it is significantly more detailed than the typical contract proposals I have seen for case rate negotiations.

One final consideration is with regard to hardware and implants and high cost drugs. The old way to negotiate these was to agree to produce an invoice for each item. The new way is to assert in the contract that the price charged is reflective of not more than x% markup and allow up to two or three spot audits per year if the payer chooses to verify and audit that the rate charged is not more than the percentage markup asserted.

As I review contracts, my take on the situation is that the majority of hospital contract analysts for the most part have not developed a written set of contracting policies or business rules, and a written contracting strategy with a checklist of those things that cause modeling woes for the business office and reimbursement problems for the finance team. As such, the process becomes reactive as opposed to responsive and proactive, and each contract is addressed with the time permitted to rush through the deal without method or data. The requisite forethought and preparation are not evident.

Managed-care reimbursement takes lots of modeling capability to slowly and deliberately model the rate so that the surprises are reduced and the risk is managed.

A fallback provision must be negotiated for when the system does not pay out as anticipated, and both parties need to have formulae that are stated within the terms of the agreement and its attachments so that some third-party arbiter or mediator can get to the essence of the original understanding between the parties in the event of a dispute without having to come up with a solution that was not part of the original understanding at the time it was signed. Unfortunately, most negotiators do not understand the terms of the contracts that they sign and therefore do not sweat the details as they should. Most also do not understand the mechanics of dispute resolution and when a dispute arises, fail to use the prescribed remedies stated in the contract and simply quit.

My apologies to all readers who have been through this and have the degree from Hard Knocks University. The vast majority of those who are unprepared for case rate and other alternatives to simple percent of charges reimbursement by their very tolerance and execution of poor contract proposals make it difficult for the rest of us to negotiate better contracts as we are considered "troublemakers," squeaky wheels, and "difficult." I take pride in it; you?

17 Understanding Capitation Performance Guarantees

Managed-care agreements for Independent Practice Associations (IPAs), Physician Hospital Organizations (PHOs), Management Services Organizations (MSOs), and Accountable Care Organizations (ACOs) are ever-increasingly being signed as full-risk, percentage-of-premium, capitated contracts. When a Health Maintenance Organization (HMO) signs an agreement with one of these groups, the capitation paid is likely to be a large sum of money each month. Therefore, contracts and agreements of this type involve heavy liability on the part of the HMO to do the necessary due diligence with the organization in regard to economic credentialing.

The last thing that an HMO wants to receive is a "Dear John" letter from the provider network comptroller postmarked from some far-off island that says, "Dear Mr. HMO-CEO, We quit. Sincerely, The IPA."

Hence, many managed-care agreements contain specific performance language with regard to the provision of covered services and the group's performance. In this short chapter, I would like to share with you some model language from contracts I have reviewed in the past so that when you see them, you will not become unnecessarily surprised.

Often, the body of the contract includes a Guarantee of Provision of Covered Services that looks like the following:

> "IPA shall establish a plan that ensures the provision and continuation of Covered Service Participants for which capitation payments have been made, and the provision and continuation of Covered Services to Participants who are confined in an inpatient facility until their discharge or expiration of covered services, in the event of group's failure to provide or continue such covered services whether such failure results from a breach of this agreement by the group, the insolvency of the group, or otherwise. The plan must be in all respects acceptable to payor in its sole and absolute discretion and may include (a) sufficient insurance to cover the expenses to be paid for such covered services; (b) provisions in provider contracts that obligate the provider to provide medically necessary covered services for the duration of the period for which capitation payments have been paid to the group and until the participant's discharge from inpatient facilities; (c) reserves in amounts sufficient to cover expenses to be paid for such medically necessary covered services; (d) letters of credit in amounts sufficient to cover expenses to be paid for such medically necessary covered services; and (e) such other arrangements that assure that the medically necessary covered services described above are provided to participants.
>
> The group shall submit its initial plan for this guarantee of covered services to payor for its approval at least twenty (20) business days prior to the effective date of this agreement.

Thereafter, the plan may not be changed in any respect without the prior written consent of the payor."

Additionally, another part of the contract or agreement is going to ask for a Guarantee of the Group's Performance:

"Prior to the effective date of this agreement, the group shall notify the payor in writing of the persons or entities that (a) are shareholders or partners of the group, (b) control any of the shareholders or partners of the group, (c) have a significant influence on the management of the group, or (d) have a significant economic interest in the group. The payor shall promptly notify the group of those persons or entities that payor desires to have guarantee the group's performance under this agreement. Group shall cause each of such persons or entities identified by payor to execute a Guarantee of Performance in favor of payor in the form of the Guarantee of Performance set forth in Exhibit A.

If there is any change in the persons or entities described in the above paragraph, the group shall immediately notify payor of such change and shall cause any additional persons or entities satisfying the criteria described above to execute Guarantee of Performance if requested by payor."

Exhibit A

Capitation Guarantee of Performance: This Guarantee of Performance is entered into as of 20__ (this "Guarantee") by John Smith, M.D., a partner in [provider organization name], a limited liability corporation incorporated under the laws of the state of [state] ("Guarantor") in favor of [HMO name], ("Payor"), and its affiliates in connection with the Group Agreement (the "Agreement") between [HMO name], and [provider organization name].

WHEREAS, Guarantor is [a partner], [a shareholder], [describe other relationship to group] that will benefit substantially and significantly from group entering into the agreement with payor.

WHEREAS, as a condition of entering into the Agreement with Group or as a condition of continuing the Agreement with Group, payor has required that Guarantor execute this guarantee in favor of payor and its affiliates.

NOW, THEREFORE, for good and valuable consideration, the receipt and sufficiency are hereby acknowledged, Guarantor hereby agrees as follows:

1. Guarantee: Guarantor hereby unconditionally and irrevocably guarantees the full, complete, and punctual financial performance of all obligations of group under the agreement. Guarantor hereby agrees that its obligations hereunder shall be unconditional, irrespective of the validity or enforceability of the agreement, the absence of any action to enforce the agreement, the waiver of any rights thereunder, or the amendment of the agreement or any other circumstance that might otherwise constitute a legal or equitable discharge or defense of a guarantor. Guarantor hereby waives diligence, a presentment, demand of performance, any right to require a proceeding first against group, protest or notice with respect to the obligations under this agreement, and all demands whatsoever. Guarantor agrees that this guarantee will not be discharged except by complete performance of the obligations contained in this guarantee.

This guarantee shall not be affected by, and shall remain in full force and effect notwithstanding, any bankruptcy, insolvency, liquidation, or reorganization of group or Guarantor.

2. Representations and Warranties: Guarantor hereby represents, warrants, and agrees as follows: (a) Guarantor has all requisite legal power to enter into this guarantee; (b) Guarantor has all requisite legal power to carry out and perform its obligations under the terms of this guarantee; and (c) this guarantee constitutes the legally valid and binding obligation of Guarantor, enforceable in accordance with its terms.
3. Governing Law: This guarantee shall be deemed to be a contract made under the laws of the state of [state] and shall for all purposes be governed by and construed in accordance with the laws of such state.

IN WITNESS WHEREOF Guarantor has executed this guarantee as of the date first above written. (Signatures and Date follow).

Would the payor be likely to change any of the foregoing language? Not hardly!

My suggestion is to have this plan ready before entering negotiations with any payor so that the time frame of the negotiation can he moved along in due course without delay. This is part of your organization development work. Do not be caught without having thought it through and having established a viable plan with reinsurance backup.

18 Considerations for Reinsurance Purchases for the Integrated Health Delivery System

In a healthcare environment that changes daily, healthcare providers are increasingly exposed to the financial risks associated with providing medical care. As you are probably aware, many medical providers have reimbursement contracts with Managed Care Organizations (MCOs) based upon capitation. Capitated contracts provide for a fixed fee paid by an MCO to a participating healthcare provider. This fixed payment is intended to cover all the expenses associated with the medical needs of the provider's patient population, and remains fixed without regard to the actual cost of medical services required by a patient. If a provider organization runs out of money before it runs out of month, that is a problem.

So what does this all mean to healthcare providers? If total patient costs fall below the capitation amount, the provider keeps any excess. But if costs exceed the capitation payments, the provider must absorb the additional costs. Capitation can work well for healthcare providers in providing routine medical care. However, when patients experience medical catastrophes, the capitated provider is exposed to a financial catastrophe.

Consider the following example:

A teenager was admitted to a hospital after being involved in a car accident, and lapsed into a coma that lasted for an extended period of time. During this period, the patient required extensive medical care and was kept alive on life-support systems. The patient's family was insured through an MCO, which had a capitation arrangement with an Independent Practice Association (IPA) and a Provider Hospital Organization (PHO).

With this arrangement, the payment for the hospital services required by the teen was limited to the fixed monthly capitation payment received by the IPA/PHO. And, because of this, the IPA/PHO had to bear significant excess expenses not paid by the MCO. The resulting out-of-pocket cost to the IPA/PHO was close to $1 million, potentially having a serious impact on the overall financial health of the IPA/PHO.

COVERING THE REMAINDER

A majority of healthcare costs stems from a small number of catastrophic cases such as the previous example. In a capitated environment, these cases critically threaten a provider financially. While some MCOs may offer provider excess coverage, providers should consider whether it is wise to seek a financial safety net from the same organization that is the source of the actual financial risk. (I have found very often that, indeed, they do charge a "convenience" surcharge; although not listed as such, it is reflected in the price quoted.) Obtaining provider excess insurance from a single insurer, distinct from the MCO, separates the financial risks assumed under a capitation arrangement from the source of the basic medical reimbursement. A single insurer also gives providers the opportunity to obtain superior coverage and reduced administrative burdens, often at lower costs than co-sponsored programs.

To protect themselves against large potential losses, healthcare providers need comprehensive insurance that is both flexible with and responsive to their unique needs. Many vendors can provide this coverage with provider excess loss insurance policies. Their coverage is customized for healthcare providers and designed to cover the excess costs associated with catastrophic medical events. Often, policies provide hospitals, physician groups, and other medical practitioners key features, such as coverage for hospitals and professional charges, flexible coinsurance provisions and deductibles, extended reporting periods, as well as additional features, that can be tailored to each provider's needs. Most healthcare providers cannot afford to be without coverage, although many in this day and age try to "run bare." It astounds me to think how many capitated providers are seeing 30 percent of actual billed charges (or less!) when capitated reports are prepared, yet they have not taken the time to investigate the possibility of reinsuring the risks they have assumed.

Catastrophic financial losses for healthcare providers under capitation arrangements can exceed $1,000,000 per patient in any given year. Provider excess loss insurance reimburses providers for catastrophic medical expenses that are not anticipated by the level of capitation payment paid by an MCO.

POLICY KEY FEATURES

Many policies includes the following key features:

Term of coverage: Many policies are issued for a twelve-month term that does not need to coincide with the term of the provider service agreement with the MCOs. If your contract is longer, you may want to purchase a longer-term policy. However, in my experience, it may pay to have the flexibility of an annually renewed/renegotiated policy that provides the opportunity to buy less coverage and pay less as you become more familiar with capitated management techniques. Most offer monthly premium rates that are guaranteed for the policy period, and can be tailored for coverage to meet the medical risk incurred by the provider under terms of its provider service agreements.

Experience rating: Newer organizations with little experience in capitated risk management as a group may be rated due to their lack of experience with capitated risk management. Although their underwriting process considers the factors developed through a pricing model, they can also consider prior claims experience in establishing premium rates. Sometimes, underwriters offer an option to structure the policy so that favorable loss experience during the policy term can be shared with the insured upon renewal.

Covered person: The typical reinsurance policy can cover commercial, Medicare, and Medicaid patients in one contract. However, premiums, deductibles, and coinsurance provisions can vary based on the unique exposures providers face in the demographics of their patient population.

Eligible expense: Reimbursable expenses reflect services and supplies provided by the insured to patients under the terms of the insured's provider service agreements with MCOs. Coverage for both hospital and professional services is available. Reimbursements can be based on reasonable and customary charges, per-diems, or the common reimbursement schedules such as McGraw-Hill, RBRVS, etc. The insurance will not reimburse amounts greater than the insured has actually paid. An actuary can often assist you in the development of internal fee schedule development for an IPA or PHO so that the MSO can manage the capitation and the risk based upon the monthly funding it receives from the health plan, rather than the full premium dollar. Eligible services such as Hospital, Physician, Transitional Care, Home Health, Pharmaceutical, and Transportation (Air Ambulance, etc.) expenses were the general expenses I found through online review.

Annual deductibles: Deductibles are generally per member, per policy year and can be tailored to meet each provider's different needs and ability to assume risk. Deductibles for hospital services can be as low as $25,000, and deductibles for professional charges can start as low as $7,500. While standard policies do not provide for a carry-forward, some offer a thirty-one-day carry-forward provision for hospital coverage. Specific deductibles I reviewed online generally ranged from $25,000 to $1 million.

Coinsurance: Often, the policies offer flexible provisions to meet each provider's needs, willingness, and ability to assume financial risk. These can frequently go all the way up to 90 percent.

Annual maximum: The typical reinsurance policy defines maximum reimbursement for each individual based on annual limits. Hospital coverage limits were, in the past, typically $1,000,000 per member, per year, and annual coverage limits for professional services can be up to $250,000. Now, with the removal of lifetime benefit maximums, this means the losses for an integrated health delivery system are also potentially greater if they accept risk transfer or partial risk.

Benefit accumulation and reporting period: Expenses for eligible claims must be incurred within the twelve-month policy term. Payment and reporting of eligible expenses by the insured can occur from three to six months after the expiration of the policy.

Technical expertise: It is best to seek out a reinsurance broker or underwriter who is skilled in capitated risk analysis. Seek out underwriters who can look beyond stereotypes of distressed industry classes or geographic and economic trends to fully evaluate the individual risk. Look for those who offer customized policies and individually tailored policy endorsements to meet the unique coverage requirements of each healthcare provider.

Stable capacity: The financial strength and stability of a reinsurance company is recognized by the ratings agencies, earning an "A" (excellent) rating from A.M. Best Company, Inc., and a claims-paying ability rating of "AI" from Standard & Poor's.

What to Consider when Buying Capitated Stop Loss/Reinsurance from a Private Insurer

Managed care, disease management, consumer-driven healthcare, and other types of medical risk management initiatives have altered traditional products and portfolios. When seeking reinsurance coverage for an integrated health delivery system or provider organization, there are several things to consider: primarily, cost and coverage.

Cost: As you are aware, you have two cost categories: (1) in your facility or contracted provider care and (2) out-of-facility or noncontracted provider care.

When you have a patient in your facility or under the care of a contracted provider, you have greater control over your costs, and you should need minimal reimbursement from stop loss, such as 50 percent or 45 percent of billed charges. However, outside your facility or noncontracted providers, you are subject to the good will of your colleagues. Our usual recommendations to clients regarding the out-of-facility care are to request a 70-percent reimbursement or even a 90-percent reimbursement of the amount that you actually pay that facility or provider.

Coverage: Many people approach the question of what level of coverage to purchase from an intuitive standpoint instead of from an analytical one. From my experience, I have learned that intuition was why you took the risk in the first place; in most cases, why not seek the backup of good analytical/actuarial science by shopping for the coverage based on facts and scientific method.

Realize that you are going to an outside insurer and you can design any level of reimbursement, within reason, and any deductible. A good actuary can assist your organization in ascertaining what your real exposure may be, so that your accounting department can determine the projected cost of doing business and the broker can then design your coverage to reimburse that cost or slightly below it.

Most providers buy coverage from brokers who specialize in this area. The benefits of a qualified broker are the following: Qualified brokers of reinsurance/stop loss assist you in the evaluation and recommendations for the appropriate way of insuring your risk.

If the broker has a substantial book of business with the various insurance companies that are available, they have more clout than you do in terms of receiving competitive pricing and resolving service issues. A "true broker" has the ability to access many different insurance markets, which increases your odds of receiving the best possible value for your premium dollar.

Selection of a broker may or may not come easily. When capitation is popular in a managed-care market, providers and identified groups will often receive many mailers or phone calls from various brokers throughout the country who are touting their ability to help them obtain this type of coverage.

Markets new to capitation may not be romanced as often, or at all. The first step, if you are in the latter category, is to go to the Internet and search for "Medical Reinsurance Underwriters." You will see a few listings.

A little secret I would like to share with you is that many providers send out their requests for quote to every broker who has ever sent them a piece of mail on the subject. Experience teaches you that this is not the best way to buy this type of coverage. Many of these brokers work with the same insurance companies, which results in multiple submissions of your risk to the same underwriters.

The underwriter at the insurance company will see that none of the brokers who are involved have control over the account and, thus, are in no position to negotiate. The underwriter will feel that he has a great chance of obtaining your account because there are so many people trying to sell his product for him. Consequently, the underwriter will not give you the best deal. He will quote his full retail rates.

I find that it is best to select a couple of the most impressive, customer service-oriented brokers and have them choose which insurance companies they plan to submit your risk to. Allow each broker to go to two or three insurers exclusively so that the underwriters will not receive multiple submissions.

I liken this to selecting a prom dress. You never want to show up at the prom and have three other ladies wearing the same gown! Make sure they do not overlap in who they submit to, do not be shy about asking, and do not be shy about telling them why! It saves everybody time and aggravation.

In selecting a broker (also often referred to as a "producer"), find a broker who has several verifiable clients that are currently being serviced by the broker. If they do not have at least ten accounts, then consider them fairly new to medical reinsurance. I find it uncanny that sometimes this "stop loss expert" knows less about the excess loss coverage than the client. Also, be careful about going to your current broker or life insurance agent, as sometimes, through their lack of experience and goal-oriented salesmanship, they may accidentally misinform you and could do more harm than good.

Select a broker/producer who is more than just an order-taker/salesman. The brokers who are technically oriented often receive more respect from underwriters and, in turn, frequently receive the best pricing for their clients.

Choose a broker who brings added value. If all the broker does is send your submission to an underwriter and comes back with a quotation that he tries to sell you, then that broker has not earned his commission. You want a broker who is going to work with you to design the coverage to your needs, aggressively negotiate your final pricing, make sure that the insurance company pays your claims in a quick and efficient manner, and deliver service after the sale, through his availability to answer questions on an ongoing basis as your questions arise, needs change, or risk exposure increases with added managed-care agreements throughout the year.

Look to do business with a true insurance broker, meaning one who has access to various underwriters in your state area. There are many brokers who only represent one or two insurance companies. When you buy coverage from these types of salespeople, you are not making a fully informed decision about what is available in the marketplace. The best way to identify this type of broker is to ask for a minimum of four quotes; if the broker cannot do this for you, he is not a "true broker." If he carries some of the same lines as other brokers, that is when you assign them to a specific underwriter so that you do not have the crossover described above.

Many providers purchase stop loss without understanding what they have bought. Look for your broker to assist and educate you in the mechanics of reinsurance and stop-loss coverages. Any broker who will not take the time to educate his client may be oriented toward intentionally taking advantage of the policyholder with unanswered questions and vague policy language.

Be careful when going to insurance societies that are outside the rating system of Standard & Poor's and A.M. Best.

Many veterans can recite a tale of woe as they describe numerous stories of dealing with such entities to purchase coverage from offshore insurance societies that were nonadmitted and not financially rated by the normal rating authorities. They recite stories of how they have been in dispute with the claims organization that was processing this European insurance society's claims. Often, they are left without proof of coverage from the company, or they are left with no verifiable documentation of coverage. They tell of tales that include the mishap of only receiving a one-page letter from their broker that listed a reimbursement formula that was unintelligible.

What to Consider when Buying Coverage from an Insurance Company

- When you buy a stop-loss/reinsurance policy from an insurance company, obtain a sample of a contract from the insurer. Your claims will not necessarily be paid in accordance with the quotation that the broker gives you. As in all contracts, respect the Entire Agreement paragraph for what it is.
- Read the entire policy and understand that any verbal, marketing, or other discussions, negotiations, arrangements, or promises are notwithstanding upon your policy unless they are included in the contract. Memorandums of understanding, or written or verbal clarifications, are not valid unless they are included in an insurance policy as a declaration, schedule, or other recognized and incorporated document.

  Have the specimen contract reviewed by your health law attorney to make sure there are not any glaring problems with it. Look for clauses that require timely filing of claims for reporting limits; some are so tight that they reduce the benefit by as much as 50 percent, although not highlighted in their proposal.
- Another technique is to give all brokers, insurers, and underwriters the same case scenario and have them explain how their company would handle the

situation. I am often surprised at the differences in approach, as well as the difference in coverage.

- Be sure the insurer is "admitted" by the Department of Insurance. Never buy from a "non-admitted" insurer; it is not worth the risk. Check on complaints and problems, and perhaps even request or obtain a copy of their annual statement filed with the Department each year.
- If a broker tries to sell you a non-admitted insurer, strongly consider working with another broker. In many states, including California, the broker is required to have you sign a disclosure statement when selling a non-admitted insurer. If the broker does not voluntarily disclose this information, get a new broker to avoid any unnecessary risk.

Non-admitted insurance companies are not subject to the financial solvency regulation and enforcement that apply to admitted insurance companies. These insurance companies do not participate in any of the insurance guarantee kinds created by admitted insurers in your state. Therefore, these funds will not pay your claims or protect your assets if the non-admitted insurance company becomes insolvent and is unable to make payments as promised.

DEALING WITH MANAGING GENERAL UNDERWRITERS (MGUs)

Because provider stop loss is a very specialized area of insurance, entrepreneurial underwriting professionals, or MGUs, convince an insurer to allow them to represent their company. The MGUs perform all the major tasks that the insurance company normally does, meaning that they receive premiums, underwrite the risk, pay the claims, and market the product. Most MGUs are very reputable organizations and are a good source to buy this coverage from, but make sure you check them out thoroughly before you buy from them.

What to Consider when Purchasing from a Managing General Underwriter

- First, determine how long has the MGU been in business. If for only three to five years, you may be taking a risk. The risk is that the MGU is collecting all your premium on behalf of the insurance company. Premium is not paid to the insurance company until the end of the year. If the MGU were to go bankrupt, commit fraud, or not have the ability to pay your claims, you might find yourself uninsured, or in court trying to get your claims paid.
- Second, ask what the finances are of the MGU. Again, think of the concerns indicated above. If it is a privately owned business controlled only by a few individuals, you are taking a risk.
- Third, buy only from in-state MGUs. If you ever have a problem, you can go to the State Department of Insurance for help. Also, if you need to take legal action, it is less expensive and easier to do this against a business in your own state versus out of state.

- Fourth, if you get a quotation from a MGU that is dramatically below the other insurers that you received quotes from, you should definitely question the viability of that coverage.
- Fifth, ask the MGU to give you a list of insurers they work with and how long they have had that relationship. Be aware of MGUs who change insurance companies every year or two. This could indicate that they are losing money for their insurers and that you are putting yourself at risk to buy from them.
- Finally, just because an insurance company has been convinced by an MGU to allow them to do business on their behalf does not mean that the MGU is a viable organization. Insurance companies are not known to be perfect in their due diligence when it comes to who they do business with. Wise advice indeed!

Proper reinsurance coverage can translate into a six- or seven-figure reimbursement to your facility, so choose wisely. The good news is that if you are a typical provider with good claims experience, you should save anywhere from 30 percent to 70 percent over what you have been paying to the HMO. Thus, you will increase your bottom-line profits and allow your negotiations with the HMO to focus solely on your capitation rates and not on stop loss. Without the knowledge of what your reinsurance terms and your actuarial projections are, never negotiate full-risk, capitated agreement dollars. You set yourself up for a loss.

19 Opportunities in Delegated Utilization Management and Claims Management for the MSO

In this chapter is a sample of a requirement often seen in managed-care capitated contracts with IPAs (Independent Physician Associations) and PHOs (Physician Hospital Organizations). Normally, these functions are assumed at the MSO (Management Services Organization) level. I have included a sample exhibit of a network provider contract for you to examine closely so that you can project costs included to be able to service the contract. In the circles that I travel, these two delegated tasks together with enrollment features are worth a significant percentage of the premium. Typically, the payors reserve upwards of 15 percent of the premium for these overhead expenses and margins.

Naturally, a fledgling MSO with immature software and systems engineering of a sophisticated information systems nature would not be able to undertake a task this grand without some expert help that might cost more than the job would pay.

CLAIMS PAYMENT RESPONSIBILITIES (DELEGATED CLAIMS PAYMENT)

Group shall administer claims for covered services rendered by represented providers in accordance with this exhibit and the terms of the agreement.

1. Group shall administer all claims for covered services in accordance with payor's claim administration standards and any other standards set forth in applicable law, including but not limited to ERISA. Group agrees to reimburse represented providers for covered services within thirty (30) days of receipt of a properly completed bill for covered services. Payor may withhold all or a portion of group's capitation payment if group repeatedly fails to reimburse represented providers on a timely basis.
2. With reasonable notice, group agrees to allow payor representatives to conduct on-site reviews of group's claims administration facilities. Such reviews shall be for the sole purpose of evaluating group's performance of

its claims administration responsibilities under this agreement, including, but not limited to, ascertaining the quality and timeliness of group's claims processing. Group agrees to correct any deficiencies detected during such reviews within sixty (60) days of payor's submission of a written report detailing such deficiencies.

3. Group shall be responsible for the production of all applicable tax reporting documents (for example, 1099s) for represented providers. Such documents shall be produced in a format and within the time frames set forth in applicable state and federal laws and regulations.
4. Group shall ensure that represented providers submit claims for covered services rendered to participants in other programs for which payor has retained claims payment responsibility directly to payor in accordance with the applicable program attachment and program requirements.
5. Group shall produce explanations of benefits for both represented providers and participants. Such explanations of benefits shall be in a format and contain data elements acceptable to payor.
6. Group shall develop and deliver training programs for represented providers that outline group's billing and reimbursement processes. Group shall make best efforts to ensure that represented providers avoid submitting claims to payor for those covered services rendered to participants for whom group has been delegated claims payment responsibility.
7. Group or its represented providers shall provide payor with encounter data on a weekly basis, showing all services provided to each participant for whom group receives capitation payments. Such encounter data shall be submitted in accordance with applicable HMO program requirements and in a format acceptable to payor. Payor may elect to withhold payment of group's compensation if group fails to submit encounter data in accordance with this agreement.

UTILIZATION MANAGEMENT (DELEGATION OF UTILIZATION MANAGEMENT)

1. Group will establish a utilization management program acceptable to payor and in accordance with NCQA (National Committee for Quality Assurance) standards. Group's utilization management program shall seek to assure that healthcare services provided to participants are medically necessary and will include, but not be limited to, the following: management of referrals between the primary care physician and specialist, prior management of inpatient services, discharge planning, major condition case management, and utilization information management.
2. Group shall prepare such periodic reports or other data as requested by payor relating to its utilization management program in a format acceptable to payor.
3. Group shall not materially modify its utilization management program without payor's prior approval.

4. Group agrees to include payor's medical director or medical director designee on group committees that are responsible for the review and continued development of group's utilization management program and other related programs.
5. Payor shall have the right to audit group's utilization management activities upon reasonable prior notice. Group shall cooperate with any such audits.
6. If payor determines that group cannot meet its utilization management obligations set forth herein, payor may elect to assume responsibility for such activities. If payor elects to assume responsibility for such activities, the parties agree to renegotiate the rams set forth in this agreement to the extent necessary, and group shall cooperate and provide to payor any information necessary to perform such activities.
7. All referrals shall be to represented providers, except where an emergency requires otherwise or in other cases where group's medical director specifically authorizes the referral. Except in an emergency, group shall require all represented providers to obtain authorization from group prior to hospital admission of any participant or outpatient surgical procedures.
8. Group or its represented providers shall provide payor with referral data on a daily or weekly basis, showing all services authorized for each participant. Such referral data shall be submitted in accordance with applicable HMO program requirements and in a format acceptable to payor may elect to withhold payment of group's compensation if group fails to submit referral data in accordance with this agreement.

I have found that the most leveraged groups are the proactive networks, are well-prepared with reinsurance and errors and omissions coverage who can agree to these terms and conditions. The main reason is that the network retains the data ownership and commensurate power to make good decisions for the group.

20 Beyond Traditional HMO and PPO Contracts

Direct Contracting with Employer-Sponsored Health Benefit ERISA Plans

Tomorrow's IPAs (Independent Physician Associations), PHOs (Physician Hospital Organizations), and MSOs (Management Services Organizations) have an opportunity to change the face of third-party reimbursement contracting as an integrated health delivery system. Employers throughout the United States are clamoring for options that do not necessarily include paying premiums to high-priced insurance companies to accept claims risk on their behalf.

The Employee Retirement Income Security Act of 1974 (ERISA)* protects the interests of participants and beneficiaries in private-sector employee benefit plans. Governmental plans and church plans generally are not subject to the law. ERISA supersedes state laws relating to employee benefit plans except for certain matters such as state insurance, banking and securities laws, and divorce property settlement orders by state courts.

An employee benefit plan may be either a pension plan (which provides retirement benefits) or a welfare benefit plan† (which provides other kinds of employee benefits such as health and disability benefits). ERISA was signed into law by President Gerald Ford on Labor Day, September 2, 1974. ERISA consists of four titles: Title I sets out specific protections of employee rights in pensions and welfare benefit plans; Title II specifies the requirements for plan qualification under the Internal Revenue Code; Title III assigns responsibilities for administration and enforcement to the Departments of Labor and Treasury; and Title IV establishes the Pension Benefit Guaranty Corporation. In this chapter, I focus only on Title I as it relates to health and welfare benefit plan concerns.

* P.L. 93-406, 88 Stat. 829 (Sept. 2, 1974). ERISA is codified at §§1001 to 1453 of Title 29, United States Code and in §§401–415 and 4972–4975 of the Internal Revenue Code.

† See ERISA §3(1), (29 U.S.C. §1002), for the different types of welfare benefit plans.

MORE ERISA PLANS THAN NCQA-ACCREDITED HMOs AND PPOs

The number of employer-sponsored, ERISA self-funded health benefit plans in the United States that have the option to design benefits to include a health travel option currently number 201,567* (Table 20.1).

This figure does not include the tens of thousands of state, county, and municipal government-sponsored, self-insured plans that also offer coverage to the teachers, firefighters, police officers, judges, state and local government office workers, and pensioner and retirees that also have the option to design benefits to include a health travel option without the interference of traditional insurance companies that may be reticent to add a health travel coverage option to plan participants.

While most of these companies represented below include Fortune 1000 companies, there is a growing trend by small to medium businesses (SMBs) of two to a hundred employees to move away from high-cost renewals charged by traditional insurance plans, and convert their plan to an ERISA self-funded health benefit plan coupled with a leased network of providers with negotiated rates and hired services of third-party administrators to re-price and process their claims, cost containment firms to negotiate out-of-network prices with providers, case managers to monitor and manage high-cost and complex cases, and health travel logistics teams to coordinate care in alternative locations away from the plan participant's home town. When coupled with an affordable reinsurance policy, self-funded plans hold the risk of a limited exposure on the cost of claims, and only pay out plan benefit dollars when a claim is incurred. This saves millions of dollars per company over traditional risk-transfer insurance policies where premiums are paid monthly regardless if claims are incurred or not.

It should come as no surprise that healthy employees boost a company's bottom line. They experience less sick time, take fewer disability days, and suffer lesser risk of premature deaths. According to the Centers for Disease Control, more than 75 percent of employers' healthcare costs and productivity losses are related to employee lifestyle choices. Self-funded health plans that couple a voluntary wellness program with their self-funded health benefit plan are currently experiencing a return on investment (ROI) of up to 7x for every dollar invested in the traditional local health delivery setting. Coupled with a health travel program that includes a comprehensive physical and age/gender-appropriate screening exams, and case management campaigns for high-cost chronic disease management and surgical interventions, that program option can easily bring in excess of quadruple those savings.

DEMYSTIFYING COVERED AND NON-COVERED SERVICES

Once an employer has decided to add a domestic or international health travel benefit option to its ERISA self-funded group health benefit plan, it needs to design its benefit option and add the details of its unique program to its Summary Plan Description (SPD). ERISA sets fiduciary standards that require employee benefit plan funds be

* Mercury Healthcare International Inc. Market Research (May 2011).

TABLE 20.1
Number of ERISA Employer-Sponsored, Self-Funded Health Benefit Plans by U.S. State

State	# of ERISA Plans	State	# of ERISA Plans
Alabama	2,331	Missouri	4,838
Alaska	291	Montana & Wyoming	738
Arizona	2,243	Nebraska	2,952
Arkansas	1,150	Nevada	865
California	14,965	New Hampshire	1,152
Colorado	3,553	New Jersey	6,964
Connecticut	3,652	New Mexico	649
Delaware	606	New York	16,341
District of Columbia	1,289	North Carolina	6,112
Florida	7,184	N & S Dakota	1,573
Georgia	5,454	Ohio	9,613
Hawaii	1,639	Oklahoma	1,956
Idaho	673	Oregon	2,436
Illinois	10,692	Pennsylvania	10,853
Indiana	5,588	Rhode Island	957
Iowa	3,021	South Carolina	2,401
Kansas	2,666	Tennessee	3,906
Kentucky	2,702	Texas	12,974
Louisiana	2,015	Utah	1,813
Maine	1,218	Vermont	959
Maryland	3,918	Virginia	6,250
Massachusetts	7,498	Washington	4,048
Michigan	6,916	West Virginia	949
Minnesota	7,030	Wisconsin	6,856
Mississippi	1,133		
		Total Plans	**201,567**

Source: Mercury Healthcare International, Inc. (May 2011) internal research.

handled prudently and in the best interests of the participants. It requires plans to inform participants of their rights under the plan and of the plan's financial status, and it gives plan participants the right to sue in federal court to recover benefits that they have earned under the plan. To be qualified for tax preferences under the Internal Revenue Code (IRC), plans must meet requirements with respect to pension plan contributions, benefits, and distributions, and there are special rules for plans that primarily benefit highly compensated employees or business owners.

ERISA specifies what the SPD must contain[*]. The SPD is the main vehicle for communicating plan rights and obligations to participants and beneficiaries. As the name suggests, it is generally a summary of the material provisions of the Plan Document, which is understandable to the average participant of the employer. However, in the context of Health and Welfare Benefit Plans, it is not uncommon for the SPD to be a combination of a complete description of the plan's terms and conditions, such as a Certificate of Coverage, and the required ERISA disclosure language.

Under Section 104(b)(1), a plan administrator must provide a summary of any material modification (SMM) in the terms of the plan as well as any change in information required to be included in the SPD[†]. This summary must be provided, in most cases, within 210 days after the close of the plan year in which the modification was adopted, and also must be furnished to the Labor Department upon request[‡]. Similar to the SPD, the materials must be written in a manner that can be understood by the average plan participant. While ERISA does not define "material modification" and does not specifically cover what changes warrant an SMM[§], courts have addressed this issue[¶]. Courts have held that plan amendments such as the establishment and elimination of benefits are material modifications[**]. However, as courts have also pointed out, not all plan amendments are material modifications[††].

Note: An insurance company's Master Contract, Certificate of Coverage, or Summary of Benefits is not a Plan Document or SPD.

An SPD must contain all of the following information:

- The Plan name.
- The Plan Sponsor/Employer's name and address.
- The Plan Sponsor's EIN[‡‡].
- The Plan Administrator's name, address, and phone number.
- Designation of any Named Fiduciaries, if other than the Plan Administrator (e.g., Claim Fiduciary).
- The Plan number for ERISA Form 5500 purposes (e.g., 501, 502, 503, etc.) (Note: each ERISA Plan should be assigned a unique number that is not used more than once.).

---

[*] *Hicks v. Fleming Cos.*, 961 F.2d 537 (5th Cir. 1992).

[†] 29 U.S.C. §1024(b)(1), ERISA §102(a); 29 U.S.C. §1022(a); 29 C.F.R. §2520.104b-3.

[‡] ERISA §104(b)(1), 29 U.S.C. §1024(b)(1); 29 C.F.R. §2520.104a-8.

[§] However, regulations provide a special rule for health plans. Subject to an exception, an SMM shall be furnished if there is a "material reduction in covered services or benefits." 29 C.F.R. §2520.104b-3.

[¶] EMPLOYEE BENEFITS LAW (Matthew Bender 2d ed.)(2000).

[**] See, for example, *Baker v. Lukens Steel Co.*, 793 F.2d 509 (3rd Cir. 1986) (elimination of an early retirement benefit option was a material modification); *American Fed'n of Grain Millers v. International Multifoods Corp.*, 1996 U.S. Dist. LEXIS 9399 (W.D.N.Y. 1996) aff'd, 116 F.3d 976 (2d Cir. 1997) (amendment to a medical plan requiring retirees to pay a portion of premiums considered a material modification).

[††] See, for example, *Hasty v. Central States, Southeast and Southwest Areas Health and Welfare Fund*, 851 F. Supp. 1250, 1256 (N.D. Ind. 1994) (amendments more specifically providing for a trustee's discretionary authority under an employee benefit plan were not a material modification because the amendments "simply clarify a power").

[‡‡] Employer Identification Number.

- Type of Plan or brief description of benefits (e.g., life, medical, dental, disability).
- The date of the end of the Plan Year for maintaining Plan's fiscal records (which may be different than the insurance policy year).
- Each Trustee's name, title, and address of principal place of business, if the Plan has a Trust.
- The name and address of the Plan's agent for service of legal process, along with a statement that service may be made on a Plan Trustee or Administrator.
- The type of Plan administration (e.g., administered by contract, insurer, or Sponsor).
- Eligibility terms (e.g., classes of eligible employees, employment waiting period, and hours per week, and the effective date of participation [e.g., next day or first of month following satisfaction of eligibility waiting period]).
- How insurer refunds (e.g., dividends, demutualization) are allocated to Participants. [Note: This is important to obtain the small Plan (<100 Participants) exception for filing Form 5500.]
- Plan Sponsor's amendment and termination rights and procedures, and what happens to Plan assets, if any, in the event of Plan termination.
- Summary of any Plan provisions governing the benefits, rights, and obligations of Participants under the Plan on termination or amendment of Plan or elimination of benefits.
- Summary of any Plan provisions governing the allocation and disposition of assets upon Plan termination.
- Claims procedures—may be furnished separately in a Certificate of Coverage, provided that the SPD explains that claims procedures are furnished automatically, without charge, in the separate document (e.g., a Certificate of Coverage), and time limits for lawsuits, if the Plan imposes them.
- A statement clearly identifying circumstances that may result in loss or denial of benefits (e.g., subrogation, Coordination of Benefits, and offset provisions).
- The standard of review for benefit decisions (we recommend consideration of granting full discretion for Plan Administrator or authorized Fiduciary to interpret Plan and make factual determinations).
- ERISA model statement of Participants' rights.
- The sources of Plan contributions, whether from employer and/or employee contributions, and the method by which they are calculated.
- Interim Summary of Material Modifications (SMMs) since SPD was adopted or last restated.
- The fact that the employer is a participating employer or a member of a controlled group.
- Whether the Plan is maintained pursuant to one or more collective bargaining agreements, and that a copy of the agreement may be obtained upon request.
- A prominent offer of assistance in a non-English language (depending on the number of participants who are literate in the same non-English language).
- Identity of insurer(s), if any.

Additional requirements for Group Health Plan SPDs:

- Detailed description of Plan provisions and exclusions (e.g., copays, deductibles, coinsurance, eligible expenses, network provider provisions, prior authorization and utilization review requirements, dollar limits, day limits, visit limits, and the extent to which new drugs, preventive care, and medical tests and devices are covered). A link to network providers should also be provided. Plan limits, exceptions, and restrictions must be conspicuous.
- Information regarding COBRA*, HIPAA†, and other federal mandates such as the Women's Health Cancer Rights Act, preexisting condition exclusion, special enrollment rules, mental health parity, coverage for adopted children, Qualified Medical Support Orders, and minimum hospital stays following childbirth.
- Name and address of health insurer(s), if any.
- Description of the role of health insurers (i.e., whether the Plan is insured by an insurance company or the insurance company is merely providing administrative services).

Recommended, but not required provisions in an SPD include the following:

- For insured arrangements, attach the Summary of Benefits provided by the insurance companies to help ensure you have provided an understandable summary of the Certificate of Coverage.
- Language such that in the event of a conflict between the Plan Document and the SPD, the Plan Document controls.

WRAP SPD DOCUMENT REQUIREMENTS

Group insurance Certificates of Insurance are typically not SPDs because they do not contain all of the language required by ERISA. An employer must prepare an ERISA "wrapper" to supplement the Certificate of Insurance. Together, the wrapper and Certificate of Insurance comprise a proper SPD.

The plan's annual report must include a detailed financial statement containing information on the plan's assets and liabilities, an actuarial statement, as well as various other information, depending on the type of plan and the number of participants. (A manual review of these reports is the only way we know of to obtain the plan participant size for ERISA plans. To date, we have found no single source that offers such statistics for public review.) Plan administrators must make copies of the annual report available at the principal office of the plan administrator and

---

* Consolidated Omnibus Budget Reconciliation Act of 1985.

† Health Insurance Portability & Accountability Act of 1996, Kennedy-Kassenbaum Bill Privacy: A bill enacted by Congress in 1996 that established a comprehensive and uniform federal standard for ensuring privacy of genetic information.

at other places as may be necessary to make pertinent information readily available to plan participants[*].

The annual report must be filed within seven months after the close of a plan year, and extensions may be available under certain circumstances[†]. The annual report is to be filed with the U.S. Department of Labor (DOL) on Form 5500[‡]. In 2006, the DOL published a rule requiring electronic filing of Form 5500 annual reports for plan years beginning on or after January 1, 2008[§].

One reason that plan administrators cite in their reluctance to jump on the health travel bandwagon is a fear of breach of fiduciary responsibility potential for lawsuits by employees.

ERISA imposes certain obligations on plan fiduciaries, persons who are generally responsible for the management and operation of employee benefit plans. ERISA Section 3(21)(A) provides that a person is a "fiduciary" to the extent that the person (1) exercises any discretionary authority or control with respect to the management of the plan or exercises any authority with respect to the management or disposition of plan assets; (2) renders investment advice for a fee or other compensation with respect to any plan asset or has any authority or responsibility to do so[¶]; or (3) has any discretionary responsibility in the administration of the plan[**]. Every plan governed by ERISA must have one or more named fiduciaries, and these fiduciaries must be named in the plan document.

Section 404(a)(1) of ERISA establishes the duties owed by a fiduciary to participants and beneficiaries of a plan. This section identifies four standards of conduct: (1) a duty of loyalty, (2) a duty of prudence, (3) a duty to diversify investments, and (4) a duty to follow plan documents to the extent that they comply with ERISA[††].

1. *Duty of loyalty.* Section 404(a)(1)(A) of ERISA requires plan fiduciaries to discharge their duties "solely in the interest of the participants and beneficiaries" and for the "exclusive purpose" of providing benefits to participants and beneficiaries and defraying reasonable expenses of administering the plan[‡‡]. The duty of loyalty applies in situations where the fiduciary is con-

* ERISA §104(b)(2), 29 U.S.C.§1024(b)(2). Under this section, other materials, such as a bargaining agreement or trust agreement affecting the plan may also be made available.

† See 29 C.F.R. §2520.104a-5.

‡ While ERISA and the Internal Revenue Code provide that other annual reports must be filed with the PBGC and the Internal Revenue Service, these reporting requirements can be satisfied by filing Form 5500 with the Labor Department.

§ 29 C.F.R. §2520.104a-2.

¶ See 29 C.F.R. §2510.3-21, which provides guidance as to when a person shall be deemed to be rendering investment advice to an employee benefit plan.

** Plan fiduciaries may include plan trustees, plan administrators, and a plan's investment managers or advisors. See Department of Labor, Fiduciary Responsibilities, available at <https://www.dol.gov/dol/topic/retirement/fiduciaryresp.htm#doltopics>.

†† ERISA §404(a)(1), 29 U.S.C. §1104(a)(1).

‡‡ This section is supplemented by Section 403(c)(1) of ERISA, which provides that the "assets of a plan shall never inure to the benefit of any employer and shall be held for the exclusive purposes of providing benefits ... and defraying reasonable expenses of administering the plan." 29 U.S.C. §1103(c)(1).

fronted with a potential conflict of interest, for instance, when a pension plan trustee has responsibilities to both the plan and the entity (such as the employer or union) sponsoring the plan[*].

However, just because an ERISA fiduciary engages in a transaction that incidentally benefits the fiduciary or a third party does not necessarily mean that a fiduciary breach has occurred[†].

In one noted case, the court in *Donovan* noted that it is not a breach of fiduciary duty if a trustee who, after careful and impartial investigation, makes a decision that while benefiting the plan, also incidentally benefits the corporation, or the fiduciaries themselves. However, fiduciary decisions must be made with an "eye single to the interests of the participants and beneficiaries"[‡]. The court articulated that the trustees have a duty to "avoid placing themselves in a position where their acts as officers and directors of the corporation will prevent their functioning with the complete loyalty to participants demanded of them as trustees of a pension plan"[§].

A plan fiduciary must also act with the "exclusive purpose" of "defraying reasonable expenses of administering the plan"[¶]. The Department of Labor has stated that, "in choosing among potential service providers, as well as in monitoring and deciding whether to retain a service provider, the trustees must objectively assess the qualifications of the service provider, the quality of the work product, and the reasonableness of the fees charged in light of the services provided"[**].

2. *Duty of prudence.* Section 404(a)(1)(B) of ERISA requires fiduciaries to act "with the care, skill, prudence, and diligence under the circumstances then prevailing that a prudent man would use in the conduct of an enterprise of a like character with like aims"[††]. When examining whether a fiduciary has violated the duty of prudence, courts typically examine the process that a fiduciary undertook in reaching a decision involving plan assets[‡‡]. If a fiduciary has taken the appropriate procedural steps, the success or failure of

[*] Craig C. Martin and Elizabeth L. Fine, *ERISA Stock Drop Cases: An Evolving Standard*, 38 J. Marshall L. Rev. 889 (2005).

[†] Ibid.

[‡] 680 F.2d at 271.

[§] Ibid.

[¶] ERISA §404(a)(1)(A)(ii), 29 U.S.C. §1104(a)(1)(A)(ii).

[**] U.S. Department of Labor, Employee Benefits Security Administration, Information Letter, July 28, 1998. <http://www.dol.gov/ebsa/regs/ILs/il072898.html>.

[††] 29 U.S.C. §1104(a)(1)(B).

[‡‡] See, for example, *GIW Industries v. Trevor, Stewart, Burton & Jacobsen*, 895 F.2d 729 (11th Cir. 1990) (investment management firm breached its duty of prudence after investing primarily in long-term, low risk government bonds and failing to take into account the liquidity needs of the plan); *Donovan v. Mazzola*, 716 F.2d 1226, 1232 (9th Cir. 1983) (court stated that test of prudence is whether "at the time they engaged in the challenged transactions, [fiduciaries] employed the appropriate methods to investigate the merits of the investment and to structure the investment").

> an investment can be irrelevant to a duty of prudence inquiry*. Regulations promulgated by the Department of Labor provide clarification as to the duty of prudence in regard to investment decisions. These regulations indicate that a fiduciary can satisfy his duty of prudence under ERISA by giving "appropriate consideration" to the facts and circumstances that the fiduciary knows or should know are relevant to an investment or investment course of action†. "Appropriate consideration" includes (a) "a determination by the fiduciary that the particular investment or investment course of action is reasonably designed, as part of the portfolio … to further the purposes of the plan, taking into consideration the risk of loss and the opportunity for gain (or other return) associated with the investment," and (b) consideration of the portfolio's composition with regard to diversification, the liquidity and current return of the portfolio relative to the anticipated cash flow requirements of the plan, and the projected return of the portfolio relative to the plan's funding objectives‡.

Most plan administrators rely upon their consulting firms to suggest with whom they should contract, and many times there are some hefty "connections" commissions paid out for steering a large employer to entities such as the Cleveland Clinic, Johns Hopkins, Mayo, Baylor, and others. The plan fiduciaries will often admit they know very little about any particular brand of hospital, its accreditation in their own country, or international hospital accreditation in countries where they have expats and pensioners living and working abroad.

They do, however, recognize the procedures and merits of the primary source verification vetting process suggested by the National Committee for Quality Assurance (www.ncqa.org), which is the organization that accredits health plans and members provider satisfaction, member satisfaction, health delivery outcomes, and plan documentation of its credentialing and privileging activities by primary source verification.

As an integrated health delivery system, you have the opportunity to approach these companies and offer your services to cut health costs, improve coordinated care delivery and outcomes, and deliver certain specialty services from new market sources that were previously unavailable.

An employer may wish to consolidate its various component benefit plans into a single plan for reporting and disclosure, thereby cutting significant administrative costs.

Group health employer benefits consultants can help benefits managers, plan administrators, and fiduciaries save the time and expense of preparing an SPD, a Form 5500, and a Summary Annual Report (SAR) for each separate component benefit plan by replacing them with one comprehensive wrap plan. If they understand the health travel benefit options, they can incorporate this valuable savings option

---

* See, for example, Unisys, 74 F.3d at 434 ("[I]f at the time an investment is made, it is an investment a prudent person would make, there is no liability if the investment later depreciates in value"). A plan fiduciary must also act with the "exclusive purpose" of "defraying reasonable expenses of administering the plan."

† See 29 C.F.R. §2550.404a-1.

‡ Ibid.

and can design a solution that is right for each plan and drive the business for certain service lines directly to your door.

The problem is that most benefits consultants are not experts in ERISA or health travel benefit options. You may have to identify spokespersons and hire a marketing consultant to make it known that you are interested in this business if you plan to attract business from elsewhere in the United States or abroad. The marketing consultant will have to understand ERISA well enough to be fluent in the sales proposal. This has been challenging, nationwide, although not impossible.

My professional opinion is that it will take several more years (I am estimating five to seven years [i.e., 2016 to 2018]) for employers to feel comfortable with international health travel as a normal option for their domestic workforce's health delivery, and a shorter time for those expatriate employees. Now in 2012, we are already witnessing unprecedented steerage, however, to domestic Centers of Excellence and also to high-value, high-performance healthcare providers.

Internal and unpublished industry statistics from the health plans that currently offer traditional coverage for expatriate employees already demonstrate a decrease in the number of expatriates returning back to the domestic healthcare providers for high-complexity care. My educated guess as to how acceptance will increase is that as these expatriates sample the care of the international marketplace and are queried for patient delight and clinical outcomes, the word will spread and the plan fiduciaries will test the waters, first with a limited number of service lines, and eventually for any services covered by the plan for which the plan participant chooses to travel for care, providing the travel can be done safely and with pricing transparency, regardless of whether the care provider is located across town, across the county, across the state, across the nation, or outside the national borders of the country in which the plan participant resides.

You will need a template contract similar to a PPO (Preferred Provider Organization) contract in order to execute an agreement with the employer plan sponsors, and these are not widely circulated in the industry. I have negotiated and executed several for clients over the past twenty years, including with Marriott, Weyerhaeuser, the International Longshoreman's Union, Tyson Foods, and others. While I would not be able to simply "send you a copy," I am available to assist if you determine that this is a direction that may bring market share to your organization.

Section IV

Appendices

Appendix A: Volunteer Committee Survey Form

GETTING EVERYBODY TO PITCH IN

In the early days of network development, very few organizations have the money to hire a team of consultants to develop the network hands-on. Most organizations opt for a consultant to lead the committees' directors toward a viable end, allowing for the network's members to pitch in with sweat equity. In order to get everyone to pitch in and create a master logistics list, a form like this one, handed out at the first meeting, may prove as helpful for you as it did for me.

COMMITTEE INTEREST SURVEY

Physician's Name: ______________________________

Telephone: ________ Office: ________ Home: ________ Cell: ________

E-mail: ______________________________

Specialty: ______________________________

I Have An Interest in Serving on the Following Committees:

- Bylaws
- Membership
- Credentialing
- Quality Assurance/Quality Improvement
- Utilization Management/Utilization Review
- Healthcare Standards /Medical Appropriateness Guidelines
- Medical Director's Position
- Risk Management
- Public Relations and Social Media Channels
- Patient Relations
- Hospital Relations
- Provider Relations
- Finance
- Budget Committee
- Contracting Committee
- Business Development
- Info Systems/Computer
- **I Have No Interest In Serving on any Committee**

Appendix B: Sample LLC Document Set

The following document is intended to apprise the person considering an investment in a new entity such as an IPA, PHO, or MSO, etc. of the material risks of the business enterprise as a limited liability corporation. When you hire an attorney or accountant to prepare this for you, these are the documents they will be preparing; this takes time and requires lots of necessary input from your Steering Committee members.

DOCUMENT PREPARATION CHECKLIST

Before an organization can prepare this document, certain types of information must be prepared for the attorney. Give a brief description of the organization, including the following information:

- Type of organization
- Services to be provided by the IDS
- Types of persons/entities to which membership is being offered
- Legal capacity description of a member
- Knowledge and experience in financial business matters enabling the member to evaluate the merits and the risk of the investment
- Members' ability to bear economic risk of the investment
- Sample copies of the
  - Articles of Incorporation
  - Bylaws
  - Membership Agreements (for example, Physician Services Agreement, PHO Participation Agreement, Management Services Agreement, Membership Agreement, etc.)
- Market value and transferability information, as applicable
- If integration is to be accomplished in stages, describe each stage
- The benefits of membership in the IDS
- Obligations of the members
- Requirements for membership in the IDS

MEMBER SUBSCRIPTION AGREEMENT

Note: This is a sample form. Federal and state law will strongly influence what disclosures should be made, and state laws can often vary. Consult experienced health law specialty legal counsel in preparing and subscription agreement to confirm its compliance with all applicable laws.

[Company Name] [Date]
To: [Company Name]

Ladies and Gentlemen:

This Member Subscription Agreement ("Subscription") sets forth the basic terms under which the undersigned ("Subscriber") will agree to purchase a membership interest ("Membership Interest") in [Company Name], (the "IDS") and participate as a member ("Member") of the IDS.

Below the IDS must customize the document with the following information:

I understand that the IDS is being formed as a [State] limited liability company.

- Type of IDS (e.g., IPA, MSO, PHO, ACO, Foundation)
- Services to be provided by the IDS
- If integration is to be accomplished in stages, describe each stage
- Benefits of membership in the IDS
- Obligations of members
- Requirements for membership in IDS

I hereby acknowledge and confirm that I have been furnished with the following information/document …

Here again, it is necessary to determine (and discuss with competent health law specialty legal counsel) what information the Subscriber needs to enable him or her to make an informed decision with respect to the purchase of a Membership Interest in the IDS. For example,

- Copies of any agreements Subscriber will he required to sign (e.g., provider services agreement, management services agreement)
- Copies of the Articles of Organization and Operating Agreement of the IDS
- Any other pertinent corporate and business information respecting the IDS

I further acknowledge and confirm as follows:

Consider including acknowledgments with respect to any one or more of the following issues, as appropriate for the nature of the IDS:

- Subscriber has read the documents supplied to him/her by the IDS and has had an opportunity to ask questions of and receive satisfactory answers from the organizers, managers, officers, directors, agents, members and/or other persons associated with the IDS, concerning the terms and conditions of Subscriber's investment.

- Subscriber understands that the MS has no history of operations and earnings and/or that his/her investment in the IDS is speculative and involves a high degree of risk which may result in the loss of the total amount of the in.
- Subscriber acknowledges that no assurances have been made regarding existing or future tax consequences that may affect him/her as a Member of the IDS.
- Subscriber has received no representations or warranties from the IDS, the organizers, the managers, any member, director, officer, or agent, and, in making this investment decision, he/she is relying solely on the information provided to him/her by the IDS and upon personal investigation.
- Subscriber acknowledges that Membership Interests are intended for purchase by persons who have such knowledge and experience in financial and business matters that they are capable of evaluating the merits and risks of the prospective investment. Subscriber (either through personal business and financial experience or through advise of a financial consultant) is capable of evaluating the investment, the IDS, the risks associated with this investment, and his/her ability to bear the economic risks of such investment.
- Subscriber acknowledgment as to certain securities issues (e.g., neither SEC nor state securities administrator has made fairness determination relating to investment, neither SEC nor state securities administrator has or will recommend or endorse any offering of Membership Interests in the IDS, Membership Interests in the IDS are/are not being registered under the Securities Act of 1934, Membership Interests are/are not being registered or otherwise qualified for sale under the "Blue Sky" laws and regulations of [State] or any other state and Membership Interests will/will not be listed on any stock or other securities exchange).

As discussed in the Appendix A, discuss with legal counsel whether this membership interest is only being offered to "Accredited Investors" and accordingly whether appropriate representations regarding the subscriber's states as an Accredited Investor should be included.

- Subscriber understands that the IDS will be a [Partnership / Corporation] for federal and state income tax purposes.

Consult with your tax counsel on whether the limited liability company is being established, for tax purposes, as partnership or as a corporation.

- Subscriber understands certain other issues relevant or unique to the IDS.

The foregoing expression of my awareness and understanding relating to the purchase of a Membership Interest in the IDS is true and accurate as of the date of this letter and shall be true and accurate as of the date of issuance of the Membership Interest. If any of the facts stated above shall change in any respect prior to the issuance of the Membership Interest, I will immediately deliver to the Managers a written statement to that effect specifying which information has changed and the reasons therefore.

I hereby covenant and agree as follows:

1. Purchase Price—I hereby agree to Purchase a Membership Interest in the IDS, at a nonrefundable purchase price of ($______) ("Purchase Price"). In consideration of the purchase and sale of such Membership Interest, I hereby agree to pay to the IDS, upon the IDS' acceptance of this Subscription, the Purchase Price by certified cashier's check made payable to the order of the IDS.
2. [Required Agreements]—I agree to enter into [name] of [agreement(s)] with the IDS.

   Describe the agreements (e.g., management services, provider services) that Subscriber (for his or her professional association or partnership) will be required to execute and Subscriber's obligations under the agreements. Consider including the following covenants/agreements:

   Agreement to pay any fees required under the agreements; and
   Consequences of termination or breach of the agreements.

3. Acceptance of Subscription; Right to Terminate Offer—The [Governing Entity] of the IDS has the right to accept or reject this Subscription, and this Subscription shall he deemed to be accepted by the IDS only when it is signed by an authorized representative of the IDS. I agree that Subscriptions need not be accepted in the order they are received, and that if within 30 days after the date of this Subscription ("Offer Period") the IDS has not received at least ______ Subscriptions, then the IDS may terminate this offering and withdraw this offer by written notice to me. Following such termination of this offering, the IDS shall have no further obligations to me. The IDS may, in its sole discretion, extend the Offer Period up to an additional 45 days upon written notice to me. I understand that the IDS will provide written notice to me of its decision to accept or reject my Subscription.
4. Closing—No later than ______ days after I receive notice of the IDS' acceptance of my Subscription, I will execute and deliver to the IDS the following:
   A. [List Required Agreements];
   B. Certified Check or Cashier's Check in the amount of the Purchase Price made payable to the order of the IDS; and
   C. Any other miscellaneous receipts, consents, or other documents deemed necessary by the IDS to effect the transactions contemplated in this Subscription.

Upon delivery of executed copies of the above agreements and the Purchase Price to the IDS ("Closing"), the IDS will issue to me a Certificate of Membership Interest. I agree that upon receipt of the Certificate of Membership Interest, I will be obligated to and agree to pay ….

Here it will be necessary to describe any fees in connection with the Required Agreements (e.g., management fee, any fees due in connection with the Physician Service Agreement).

5. Indemnification—I understand that the IDS and the managers/officers/directors will rely upon all representations, warranties, and covenants in this letter, and therefore I hereby agree to indemnify and hold harmless the IDS and each member, manager, director, officer, employee, and/or agent thereof from and against any and all loss, damage, or liability due to or arising out of the breach of any such representation, warranty, or covenant. All representations, warranties, and covenants contained in this Subscription, and the indemnification contained in this paragraph 5, will survive the acceptance of this Subscription and my admission as a Member of the IDS.
6. Confidentiality—I agree to hold in confidence and not disclose to any other parties any information or data concerning the IDS supplied to me, except to my lawyers, accountants, employees, and other agents, as necessary to evaluate this investment. I agree that this confidentiality obligation will be binding upon me and will continue after the termination of this Subscription, whether or not all of the matters contemplated in this Subscription are consummated.
7. Structure and Operation—I understand certain aspects of the planned structure and operation of the IDS to be as follows:

Describe, in reasonable detail, aspects of the IDS structure and operation that might impact Subscriber's investment decision. For example:

- *Capitalization (e.g., initial capital contribution, mandatory capital contribution, optional capital contribution)*
- *Restrictions on transfer or assignment of Membership interest*
- *Disqualification of Member*
- *Withdrawal by Member*
- *IDS' purchase rights (e.g., price and term of pay-out) upon Member disqualification or withdrawal*
- *Circumstances under which Subscriber would forfeit all or a significant portion of his or her Membership Interest.*

8. Articles and Regulations [Operating Agreement]—I understand and agree that, in addition to the provisions of this Subscription, my Membership Interest will be subject to and I agree to be bound by the provisions of the Articles of Organization and [Regulations Operating Agreement].

Check with legal counsel to include a description of the appropriate governing documents under the applicable state's limited liability company act.

Date: ________ SUBSCRIBER: ______________________________________

Signature of Subscriber: ______________________________

Printed Name: ______________________________

Address: ______________________________

ACCEPTED BY: ______________________________

[Business name] ______________________________

By: ______________________________

Appendix C: How to Hire the Right Consultants

Few provider organizations have been formed without the assistance of consultants and contractors to help the group form, shorten learning curves, define roles, products, manage change, and bring in expertise.

Rather than get all preachy about the perils of hiring incompetent help that will gladly take your money, produce mediocre deliverables, if any, and cause you lots of frustration and risk sounding preachy or bad-mouthing would-be colleagues, I have decided to take the high road and give you a valuable tool and suggestions on how to select and work with consultants. As a consultant myself, why would I do this? Because I am not threatened by those who are capable. And *you* face more peril by engaging those who are incompetent, not me. In fact, if and when you engage them, there will be opportunity for others like me to come and fix what they did.

As I see it, there are more than 7,500 acute-care inpatient hospitals out there. Most will build a "something," whether it is an IPA, PHO, MSO, ACO, or IDS. If they already have one, they may read this book and decide they need to change something. Net result: 7,500+ potential clients and nobody can serve all of them.

Each of these hospitals has a medical staff who will likely engage in a "something," whether it is a single-specialty IPA, a multi-specialty IPA, a PHO, an ACO, an ABC, or an XYZ. It will not matter what you will call it; what will matter is what it does and how well it serves a market need to thrive in the business of healthcare. When I look at confirmed "joiners," all I have to do is look at how many physicians there are in PHCS and MultiPlan, the largest independent PPO (Preferred Provider Organization) networks in the United States. More than 650,000 physicians participate. Given that the antitrust rules set a framework of how many physicians in a relevant geographic market or a metropolitan statistical area can be in one of these "thingies," it means there are roughly 196,000 provider organizations to be formed over the next few years. Those who are already in one will splinter off into new ones when the alignment objectives are not met, the politics do not align, and still others will want to change, update, revise, and perhaps set new cultures.

But wait! Physicians and hospitals are not the only provider organizations out there. I have built provider organizations of orthotics and prosthetics providers, physical therapists and occupational therapists, PHOs out of ambulatory medical centers and their medical staff, of which there are about 9,000 in the United States, complementary and alternative medicine providers, chiropractors, and others.

So suffice it to say, there is no need for me to hoard knowledge. If you read this book, you will be a better shopper for consulting services and a better-prepared, easier-to-work-with client. You will be a better critical thinker because you will have learned things in this book that will empower you, prepare you, and give you checklists, templates, tools, and techniques to prepare you for what lies ahead.

TERMINOLOGY

First, I would like to clarify a point about terminology and the use of labels. In my opinion, there are two types of assistance you can find: *consultants* and *contractors.*

Consultants can be internal or external. This means they can be hired on as regular employees, such as an in-house attorney, CISSP, database administrator, accountant, or other subject matter expert. They dedicate their working hours to your project only, and they serve as the go-to experts who are essentially always on call to you and your organization.

Both external and internal consultants share the characteristics of helping their clients address problems and improve business and organization results; they have a passion for the wisdom and expertise they bring to the organization, and they have the ability to galvanize clients into action. Yet those of us who have spent years in both roles know that there are significant differences in perspectives, challenges, and requirements. External consultants are often brought in because they bring wisdom, objectivity, and expertise to the organization. They are seen as gurus or saviors bringing wise counsel. Internal consultants have expertise, but it is valued differently as an organization insider.

The *external consultant* is usually viewed as having higher levels of expertise, experience, and credibility, especially if he or she is published, credentialed, and well known. This gives the external consultant more influence and buy-in from senior-level executives who may prefer to hear from outsiders on occasion.

Paying for services also implies that the output is better or more valued—but not always. In addition to these perceived advantages, externals are frequently assumed to be more up-to-date on the newest business thinking and new ways of working, and it is assumed that they will bring the added value of a broader base of experience. This one point has been woefully proven inaccurate in too many cases. While this broader experience, if it is real, can provide benchmarking and best practices as well as insights into potential pitfalls learned from other clients, that does not happen if the consultant is just somebody with a great sales pitch who happens to be between jobs. Those who hire externals assume that they will receive value in the form of deliverables in exchange for money and time invested by clients. They assume that they are purchasing outsider objectivity and ability to give tough feedback or to ask the difficult questions. Hopefully the questions I supply below will help you to make fewer assumptions and ask for more facts and proof before engagements are inked.

Internal consultants are often incorrectly and inappropriately limited by perceptions and position in the organization. Their value is often pegged at a different level and with a different expectation. They develop valuable in-depth knowledge of the business that the external consultant can rarely replicate, and often serve the organization and the management with day-to-day advice. What happens too often is that jealousy ensues and rather than lean on the outside consultant for the value the right consultant can bring, they fear they will be "outed" about something they may have one incorrectly, not well enough, or "whatever." They often see an outside consultant as a threat. Much of this comes from how the introduction of the idea of using an outsider is delivered and framed. Many, if they view the consultant as their equal, are afraid they will be replaced by the consultant from the outside, which is generally the furthest thing from both the consultant's and the company's minds. However, I caution that if the external is one of those "consultants" who is between jobs (hint: has no staff, no overhead, no investment in being a consultant), that threat is more real than you may wish to acknowledge for both the internal consultant and the external consultant who may actually charge you for the time he or she spends skillfully making the internal look bad.

As for internal consultants, their in-depth knowledge makes them particularly valuable on sensitive implementation of strategic change projects or culture transformation initiatives, managing processes or projects, and integrating or leveraging initiatives across the organization. Unfortunately, many organizations do not recognize the value of a strong and competent internal consulting function, so they hire less-experienced or less-competent junior consultants and place them in uninfluential lower positions in the hierarchy. This is something I have frequently noted in IPAs, PHOs, and MSOs over the years. Often, the role is filled by someone's office manager who wants a new job, or by one of the former externals who in many cases turns out to be an expensive corporate liability.

Then there are the contractors. *Contractors* are often inappropriately referred to as consultants. They do not provide specialized knowledge or advice. They do a job. They fill a role, often part time, and carry out tasks. Most likely, they are actually part-time employees that the provider organization has engaged, inappropriately in many cases, as independent contractors to get around payroll and workers' compensation liability. See Table C.1 for a comparison matrix of when and how best to use each.

The general rule is that an individual is an independent contractor if the payor has the right to control or direct only the result of the work and not what will be done and how it will be done. If an employer–employee relationship exists (regardless of what the relationship is called), one is not an independent contractor and one's earnings are generally not subject to Medicare and Social Security Taxes for Self-Employed.

One way you might approach this concern is called the "ABC" test, which is used by almost two-thirds of the states (not including Texas) in determining whether workers are employees or independent contractors for state unemployment tax purposes. In order to be considered an independent contractor, a worker must meet three separate criteria (some states require only that two criteria be met):

1. The worker is free from control or direction in the performance of the work.
2. The work is done outside the usual course of the company's business and is done off the premises of the business.
3. The worker is customarily engaged in an independent trade, occupation, profession, or business.

IRS Independent Contractor Test

The IRS (Internal Revenue Service) formerly used what has become known as the "Twenty Factor" test. Under pressure from Congress and from representatives of labor and business, it has recently attempted to simplify and refine the test, consolidating the twenty factors into eleven main tests, and organizing them into three main groups: behavioral control, financial control, and the type of relationship of

TABLE C.1
Consultant Contractor Matrix

Use External Consultants When:	Use Internal Consultants When:	Use Contractors When:
To support development of strategy or facilitate corporate-wide initiatives or key priorities	To support implementation of strategic priority, or intervention as an operational focus	To do specific tasks
Do not have internal expertise	Have the internal expertise	Neither internal nor external expertise, simply task-oriented skills to get a job done
Deep expertise is needed	Broad generalist knowledge is needed	Task-oriented skills are needed without tools supplied by the contracting entity (e.g., data entry, filing, processing applications, etc.)
An outside, neutral perspective is important	Knowledge of the organization and business is critical	Knowledge of how to complete the task is adequate.
New, risky alternatives need validation from an outside expert	Speaking the jargon or the language of the organization and the culture is important	Not concerned with politics, jargon, or culture, works independently, often offsite.
Internal does not have status, power, or authority to influence senior management or the culture	A sensitive insider who knows the issues is needed	Always an outsider, always casual, on call, as needed to help out
CEO, president, or senior leaders need coach, guide, or objective-sounding board	Need to sustain a long-term initiative where internal ownership is important	Works without any authority in a non-advisory role
Initiative justifies the expense	Cost is a factor	Paid by the hour or the project, plus about 25% more than same wage for an employee
Project has defined boundaries or limits	Follow-up and quick access is needed	Work is sporadic and a full-time or part-time equivalent is not always required

Source: Courtesy of Maria K Todd, Mercury Healthcare Advisory Group, Inc.

the parties. Those factors appear below, along with comments regarding each one. (Source: IRS Publication 15-A, 2010 edition, page 6; available for downloading from <http://www.irs.gov/pub/irs-pdf/p15a.pdf> (PDF). Another good IRS resource for understanding the independent contractor tests is at <http://www.irs.gov/businesses/small/article/0,,id=99921,00.html>.)

Behavioral Control

Facts that show whether the business has a right to direct and control how the worker does the task for which the worker is hired include the type and degree of

1. *Instructions the business gives the worker.* An employee is generally subject to the business' instructions about when, where, and how to work. All of the following are examples of types of instructions about how to do work:
   a. When and where to do the work
   b. What tools or equipment to use
   c. What workers to hire or to assist with the work
   d. Where to purchase supplies and services
   e. What work must be performed by a specified individual
   f. What order or sequence to follow

The amount of instruction needed varies among different jobs. Even if no instructions are given, sufficient behavioral control may exist if the employer has the right to control how the work results are achieved. A business may lack the knowledge to instruct some highly specialized professionals; in other cases, the task may require little or no instruction. The key consideration is whether the business has retained the right to control the details of a worker's performance or instead has given up that right.

2. *Training the business gives the worker.* An employee may be trained to perform services in a particular manner. Independent contractors ordinarily use their own methods.

Financial Control

Facts that show whether the business has a right to control the business aspects of the worker's job include

3. *The extent to which the worker has unreimbursed business expenses.* Independent contractors are more likely to have unreimbursed expenses than are employees. Fixed ongoing costs that are incurred regardless of whether work is currently being performed are especially important. However, employees may also incur unreimbursed expenses in connection with the services they perform for their business.
4. *The extent of the worker's investment.* An employee usually has no investment in the work other than his or her own time. An independent contractor often has a significant investment in the facilities he or she uses in performing services for someone else. However, a significant investment is not necessary for independent contractor status.
5. *The extent to which the worker makes services available to the relevant market.* An independent contractor is generally free to seek out business opportunities. Independent contractors often advertise, maintain a visible business location, and are available to work in the relevant market.
6. *How the business pays the worker.* An employee is generally guaranteed a regular wage amount for an hourly, weekly, or other period of time. This usually indicates that a worker is an employee, even when the wage or salary is supplemented by a commission. An independent contractor is usually

paid by a flat fee for the job. However, it is common in some professions, such as law, to pay independent contractors hourly.

7. *The extent to which the worker can realize a profit or loss.* Because an employer usually provides employees a workplace, tools, materials, equipment, and supplies needed for the work, and generally pays the costs of doing business, employees do not have an opportunity to make a profit or loss. An independent contractor can make a profit or loss.

Type of Relationship

Facts that show the parties' type of relationship include

8. *Written contracts describing the relationship the parties intended to create.* This is probably the least important of the criteria, as what really matters is the nature of the underlying work relationship, not what the parties choose to call it. However, in close cases, the written contract can make a difference.
9. *Whether the business provides the worker with employee-type benefits, such as insurance, a pension plan, vacation pay, or sick pay.* The power to grant benefits carries with it the power to take away those benefits, which is a power generally exercised by employers over employees. A true independent contractor will finance his or her own benefits out of the overall profits of the enterprise.
10. *The permanency of the relationship.* If the company engages a worker with the expectation that the relationship will continue indefinitely, rather than for a specific project or period, this is generally considered evidence that the intent was to create an employer–employee relationship.
11. *The extent to which services performed by the worker are a key aspect of the regular business of the company.* If a worker provides services that are a key aspect of the company's regular business activity, it is more likely that the company will have the right to direct and control his or her activities. For example, if a law firm hires an attorney, it is likely that it will present the attorney's work as its own and would have the right to control or direct that work. This would indicate an employer–employee relationship.

There is a "safe harbor" rule in Section 530(a) of the Revenue Act of 1978 that sometimes allows companies to classify workers in close cases as independent contractors, even if they might be considered employees under the IRS eleven-factor test shown here, as long as such a classification is consistent with the industry practice for such workers, or a previous IRS audit has found that such workers are not employees, or an IRS ruling or opinion letter supports the classification in question and the worker has been treated all along as an independent contractor.

Do not underestimate the difficulty in applying these standards to specific individuals performing services. In doubtful cases, always consult a knowledgeable labor and employment law attorney.

So now that we have terminology out of the way, let's move on to the next.

KNOW WHAT YOU WANT FROM THE CONSULTANT

If you call my office and ask one of the five secretaries I employ to speak with me, they will ask you to describe the nature of your call. There are four other secretaries to cover the rest of the consultants' calls. I do not say that to be snooty; I say it because chances are that you will not always get me on the first call, and you will not be calling my cell phone directly.

From that very first call, however, I will expect a certain preparation of you before I can ethically agree to take your call, learn about your organization, or determine whether or not I can accept you as a client. I may not have the time or capacity to take on another client and provide quality service, I may be conflicted with a competitor engagement, or I may be the wrong consultant for the job.

As such, I am going to ask some probing questions. This can go one of two ways: either you can supply me with the information I need in a fact-based dossier, or I can charge you for the time to ask you each question and discover all this information and write it down for myself. Personally, I prefer if you do it and supply the information so that I do not miss anything. But, it's not just me. You need to do a similar exercise with accountants, IT consultants, lawyers, marketing consultants, and the like.

If you have an internal expert already on board, then also tell us what they can supply to augment the engagement you have in mind. Otherwise, I have to assume there is no augmentation and I must supply any assistance as part of the proposal.

Next, please understand that busy consultants in niche practices rarely have time to draw up fancy, long, drawn-out, step-by-step proposals. Those who are not working have all the time in the world and lack the experience to know that many clients shop consultants, get proposals that take between sixteen and thirty hours to produce, and then attempt to do the job on the their own. (Beginners suffer this as a rite of passage.) As a result, unless you are hiring a big firm, which is highly unlikely (not, for many reasons, the least of which is cost), few have the necessary expertise without sending an entire group of consultants to your site to do what a "boutique" firm with the right experts can accomplish less expensively and more effectively.

Furthermore, the cost goes up substantially if we have to read your mind <smirk>. So here is what you need to prepare before you pick up the phone or write me or my colleagues an e-mail.

- **Who you are:**
  - The makeup of your group, number, specialty, location, and if you are for for-profit or not-for-profit.
  - What do you call your group? IPA, PHO, MSO, ACO, Thingie, WhatsIt, or Whatchamacallit?
  - Who is on the Board and the point of contact, name, phone(s), e-mail addresses, secretary or assistant.
- **What is the nature of the project:**
  - What is the problem or reason for the project or change?
  - Why has the problem surfaced?
  - What are the implications of doing nothing?

- Who is the real/end client, and what is their level of "buy-in" to the proposed project?
- Who are the end consumers (people who will be affected), and what is their support for the project?
- What trade-offs will be to be made to deliver the final change or deliverable? (What will have to be given up?)
- How will things be better or different once the project is complete?
- What concerns do you have about the project?
- What is the background? What has already been tried?
- How will you know when it has been successful?

- **Clarification:**
  - What is the reason for the current situation?
  - What evidence do you have to indicate that there is a problem?
  - How sure are you as to the cause of the problem?
  - Do you have any concerns about factors that might impact the project?
  - What concerns would users or organization members of the change voice about what will happen? (Including internal experts, if any!)
  - Are there any side effects that could arise from undertaking the project?
  - Who else is involved in the change? Do you have their unconditional support? Financial support?
  - Who will any change have an impact on?
  - Who can stop the change from being successful?
  - What are the unspoken or shadow issues that might cause the project or change to fail?
- **Create:**
  - What constraints are there on any proposed solution?
  - What are the criteria for a successful solution?
  - Is there anything we cannot do?
  - What is the budget and timescale?
  - What have you thought of already?
  - What has been tried before?
  - What risks are you prepared to take?
  - What flexibility is there in any proposed solution?
  - How will you know when you see the right solution?
- **Change:**
  - Who will be impacted by the change?
  - What will their response be?
  - What methods will you be prepared to use to implement change? (controlled to empathetic)
  - Will we have the necessary power to effect a successful outcome?
  - What other changes are taking place that will impact our project?
  - Do you have a standard engagement/deployment process that must be followed?
  - Are there any aspects of the change that we will not be managing?
  - How brutal are you prepared to be to make the change happen?
  - Where is the power to effect change held?

- Do you appreciate the full cost involved in effecting this change successfully? (Time, cash, and political issues?)
- Have you segmented those people who will and will not resist, and who the key influencers might be?

- **Confirm:**
  - How important is it for measurement to take place?
  - Are you prepared to pay for the measurement to take place?
  - Will you use quantitative or qualitative measures?
  - Who will you measure the organization's member's buy-in to the changes?
  - Who will undertake measurement?
  - What measures have you used in the past?
  - How will you measure our performance?
  - For how long will measurements continue?
- **Continue:**
  - How long do you want the change to last?
  - Have you tried this before? Did it last? If not, why?
  - What can we do to help ensure that the change will last?
  - Are you prepared to invest in things that will make it last?
  - Do you have the resources in place to support any change/the project?
  - Are responsibilities defined to maintain the change once it is complete?
  - Is there anyone who will try to eradicate the change once it is complete?
- **Close:**
  - What does "good" look like?
  - Once the change is complete, what differentiated value will we have added?
  - What are the learning objectives of the exercise you want to realize?
  - How can this learning be used elsewhere?
  - What can we do to ensure that you are not dependent on us once the change is complete?
  - What do we have to do for you to recommend us to others?
  - What else might we be able to help you with?

As you can see, these questions are practical; they will save your time and your money if you prepare the answers in advance of the first call, attach your business plan Executive Summary, and disclose what your budget limitations might be to complete the project.

QUALIFICATIONS

There is nothing here about management consultants and certifications as a consultant. In healthcare there are no real certifications that address this area of expertise. There are actually two parts to this. One is consulting skill or technique:

- How to listen
- How to teach

- How to present and persuade
- How to document notes and findings
- How to develop action plans
- How to motivate change in groups
- How to quote projects
- How to develop proposals
- How to sell consulting
- Ethical consulting conduct

The other is subject matter expertise. This expertise comes not from book learning, because book learning is usually just the "what." Most MBA-HA and MHA Programs do not go into IPA, PHO, MSO, ACO organizational development, operations, or management. Most barely go into managed-care contract evaluation, negotiation skills, or managed-care contract financial modeling. Few go into predictive health analytics, evidence-based medicine, healthcare marketing, and only very lightly into statistics, and even less into epidemiology implications that allow one to estimate or determine the size of a viable market.

Therefore, it will be a very rare recent graduate that can hit the ground with instant productivity. That does not mean that you could not bring some of this rising new talent on as an intern, for an unpaid internship, but the labor laws do not permit interns to work on day-to-day activities. They must be engaged to work on a specific project. They create deliverables for a grade.

There is no organization that currently vets or has experts qualified by any certification. The Healthcare Financial Management Association used to have a managed-care certification, but they abandoned it just a few years ago. So really, there are few credentials to ask for in this particular subdomain of healthcare business development.

Education and Training

Education and training are not, in and of themselves, expertise. They are the foundation. Fundamentals are not expertise. *Expertise* is a word used to describe someone with extensive knowledge or ability based on research, experience, or occupation and in a particular area of study. As such, training and education are on a different part of the scale from expertise.

Experts have prolonged or intense experience through practice and education in a particular field. In specific fields such as the development or organization change facilitator for an IPA, PHO, or MSO, the definition of an expert must be established by consensus because there is no professional or academic qualification for them to be accepted as an expert. Few who were successful building these organizations in the past and dealing with capitation in the 1990s are still in the business.

An expert can be, by virtue of credential, training, education, profession, publication, or experience, believed to have special knowledge of a subject beyond that of the average person, sufficient that others may officially (and legally) rely upon the individual's opinion.

As such, the consultants you hire should have professional liability insurance for errors and omissions if you plan to hire them and rely upon their guidance for things such as accounting compliance, interpretation of laws and regulations, etc. Expertise consists of those characteristics, skills, and knowledge of a person (that is, expert) or of a system that distinguish experts from novices and less-experienced people. Therefore, you will want to create a question list before you engage expensive experts. An important feature of expert performance seems to be the way in which experts are able to rapidly retrieve complex configurations of information from long-term memory. If they have not been in the industry long term, that will be difficult to achieve. They recognize situations because they have meaning. When I walk into a foreign hospital, the word on the street is that in fifteen minutes, I have determined what is going on around me (good or bad), what needs attention, and what is in trouble. The other four hours are merely confirmation.

Ericsson and Staszewski[*] confront "the paradox of expertise and claims that people not only acquire content knowledge as they practice cognitive skills, they also develop mechanisms that enable them to use a large and familiar knowledge base efficiently." If you think about it, the situation is the same for a long-established and experienced medical specialist and a PGY 3 resident.

Those who have studied problem solving suggest that "while the schemas of both novices and experts are activated by the same features of a problem statement, the experts' schemas contain more procedural knowledge, which aid in determining which principle to apply, and novices' schemas contain mostly declarative knowledge which do not aid in determining methods for solution"[†].

I do not want to turn this into a Ph.D. thesis on how to evaluate consultants. In this particular niche domain, there are very few experts, many who would like to be, and no place to endeavor advanced study or achieve certification. Finding a trusted advisor is not something achieved on the Internet in a few short hours. Take your time. The right counsel will be far and away one of the most important decisions you make as you build or redesign your organization. Choose wisely.

[*] Ericsson, Anders K., and Stasewski, James J. (1989). Chapter 9: Skilled Memory and Expertise: Mechanisms of Exceptional Performance. In David Klahr and Kenneth Kotovsky, *Complex Information Processing: The Impact of Herbert A. Simon*. Hillsdale NJ: Lawrence Erlbaum Associates.

[†] Chi, M. T. H., Glasser, R., and Rees, E. (1982). Expertise in problem solving. In R. J. Sternberg (Ed.), *Advances in the Psychology of Human Intelligence*. (Vol. 1, pp. 7–75). Hillsdale, NJ: Lawrence Erlbaum Associates.

Index

Q

R

W